The Nationalization of the Venezuelan Oil Industry

The Nationalization of the Venezuelan Oil Industry

From Technocratic Success to Political Failure

Gustavo Coronel

LexingtonBooks
D.C. Heath and Company
Lexington, Massachusetts
Toronto

4115

Library of Congress Cataloging in Publication Data

Coronel, Gustavo.
  The nationalization of the Venezuelan oil industry.

  Bibliography: p.
  Includes index.
  1. Petroleum industry and trade—Government ownership—Venezuela.
I. Title.
HD9574.V42C63   1983   338.2'7282'0987   83–47894
ISBN 0–669–06763–6

Published simultaneously in Canada

Printed in the United States of America

International Standard Book Number: 0–669–06763–6

Library of Congress Catalog Card Number: 83–47894

To the men and women
of the Venezuelan oil industry

Contents

Contents

Figures and Tables

Preface and Acknowledgments

This book was written between February and December 1982 at the Center for International Affairs of Harvard University. Being a fellow at Harvard gave me access to the excellent research facilities of the university and to the valuable help and advice of scholars such as Raymond Vernon, Jorge Dominguez, Terry Karl, Richard Mallon, Benjamin Brown, and Robert Putnam. The friendship and constant encouragement I received from Harvard Fellow and Venezuelan countryman Pedro Pick were extremely important to the progress of this work. My friends Jorge Olavarría, Alberto Quirós, Francisco Gonzalez, and Rafael Tudela gave me much-needed moral support. The Polar Foundation of Caracas generously provided funds to cover a portion of my expenses at the center. Mrs. Barbara Sindriglis and Mrs. Eva Morvay efficiently handled most of the typing.

The solidarity shown by my wife Marianela and by my three children Gustavo, Corina, and Ana with my decision to change the glamorous and well-paid corporate life for the more tranquil and modest academic environment gave me the peace of mind which is highly conducive to writing.

This is not an academic work. It is more a personal testimony, an attempt at putting down on paper experiences that could be useful to those readers who wish to obtain an insight into the nature of the changes which can take place both in people and in large corporations when nationalization takes place.

It can best be described as a case study.

List of Terms

Acción Democrática *(AD)*. A Social Democrat political party.

AGROPET. Agrupación de Orientación Petrolera, the association of oil industry technical and managerial staff.

Bariven. The materials acquisition company of PDVSA.

BCV. The Venezuelan Central Bank.

BTV. The Venezuelan Worker's Bank.

CADAFE. The largest of the Venezuelan state-owned electricity companies.

COPEI. A Christian Democrat political party, presently in power.

CVG. Venezuelan Guayana development corporation, a huge state-owned conglomerate trying to develop Southern Venezuela.

CVP. Corporación Venezolana del Petróleo, the Venezuelan Petroleum Corporation, the national oil company created in 1960 merged into Corpoven in 1978.

Fedecamaras. The Federation of Chambers of Commerce and Production, the most influential business association in the country.

INTEVEP. The Venezuelan Institute of Petroleum and Petrochemical Technology, the research and development company of PDVSA.

Lagoven, Maraven, Meneven, Corpoven. The four operating oil companies of the Venezuelan nationalized oil industry, under Petróleos de Venezuela (PDVSA).

MAS. A medium-sized socialist political party.

Ministry of Energy and Mines. The ministry having jurisdiction over the hydrocarbons, minerals, and, in general, over the energy resources of Venezuela. Previously known as Ministry of Mines and Hydrocarbons.

MIR. A small Marxist political party.

MEP. A small socialist political party.

OPEC. The Organization of Petroleum Exporting Countries.

PCV. The very small Venezuelan Communist party.

Pequiven. The Petrochemical Corporation of PDVSA.

Petróleos de Venezuela *(PDVSA)*. The holding company of the Venezuelan oil industry. Occasionally called Petroven by the U.S. press.

Pro-Venezuela. A nationalist association of businesspeople and intellectuals.

URD. Union Republicana Democrática. A small political party.

Part I
The Concession Years

1

Origins and Early History, 1878-1940

It is now believed that Venezuelan (oil) production has passed its peak.
—C.W. Hamilton[1]

When Hamilton made his prediction, Venezuelan oil production was on the order of 370,000 barrels per day. It would take 40 more years for Venezuelan oil production to pass its peak, and the volume would be exactly 10 times higher than was attained in 1930. In 1970 Venezuela averaged 3,708,000 barrels per day; only then did it start to decline.

In 65 years of oil activity in the country (1917–1982), cumulative production was almost 39 billion barrels of oil, 90 percent of which was exported. The story of this often fascinating development has been told only partially in books and trade journals. Perhaps the best of the books dealing with the historical development of the Venezuelan oil industry is Luis Vallenilla's *Auge, Declinación y Porvenir del Petróleo Venezolano,* published in 1973.[2] Rather than a detailed historical reconstruction, a background to our main period of interest, which starts at about 1970, is offered here.

State Ownership of Hydrocarbons

The concept of state ownership of hydrocarbons reservoirs in Venezuela originates in Spanish law. As early as the twelfth century, the King of Castille had already reserved the mines located in his land for the exclusive use of the crown. In his book on the chronology of Venezuelan petroleum, A. Martínez describes in detail the sequence of historical precedents on which the state ownership of the subsoil in Venezuela was based.[3] He cites the following events:

December 9, 1526   King Charles I of Spain declares that all mineral deposits to be found in lands to be discovered by Spain will belong to the crown.

1559   Phillip II, in Valladolid, issues an edict incorporating all mineral deposits to the crown.

3

August 22, 1584	Phillip II issues, in San Lorenzo, an updated version of the concept of crown ownership of mineral deposits.
1602	The governors of newly discovered lands are authorized, through the laws of the Indies, to apply the Spanish mineral deposits laws in their territories.
May 18, 1680	Charles II issues a compilation of the laws of the Indies confirming previous decisions and applying the San Lorenzo Edict to America.
May 22, 1783	Charles III issues the mining edict for New Spain, in Aranjuez. Clause 22, Sixth Title, mentions specifically "any other fossils . . . bitumens of juices of the earth" and defines all mines as the property of the crown.
April 27, 1784	The mining edicts of Aranjuez are applied to the Venezuelan territory.
October 24, 1829	Simón Bolívar issues a mining decree, in Quito, in which the property of mines "of any kind" is restricted to the national government.
April 24, 1832	The Venezuelan Congress ratifies the Quito decree of Simón Bolívar and also accepts the validity of the Aranjuez mining edicts.
March 15, 1854	The Venezuelan first mining code is issued following the Aranjuez edict. A year later, in January 1855, President J.G. Monagas issues the regulating clauses of that code. Article 2 establishes the need to obtain concessions in order to work the mines property of the nation.
April 13, 1864	The federal government transfers to the states the property of oil and other fossil fuels.
August 24, 1865	The president of Zulia State gives Camilo Ferrand, a U.S. citizen, a concession to drill for and exploit oil in any portion of the state but this concession, never developed, expires within the year.[4]

By the 1870s, therefore, the concept of state ownership of mineral deposits had been firmly established in Venezuela and would not change.

The First Oil Concessions

In 1878 the president of the Great State of the Andes[5] gave Mrs. Dolores Pulido a concession in the area of Rubio to explore and produce oil. Mrs. Pulido's husband Manuel Antonio Pulido formed a company together with J.A. Baldó, R. Maldonado, C. Gonzalez, J. Villafañe, and P.R. Rincones to conduct activities at the site of "La Alquitrana." The company, called *Petrolia,* became the first Venezuelan oil corporation. P.R. Rincones traveled to Pennsylvania, where he studied the latest techniques of drilling for oil. Rincones acquired a drilling unit and shipped it to Venezuela, where the new company started to drill for oil. It was not until 1883, however, that the first oil well, Eureka-1, was successfully completed by Petrolia. The first totally Venezuelan oil company was in operation until 1934, when it closed down after supplying crude oil and products to the western Venezuelan regions for about 50 years.

During the last decade of the nineteenth century, Venezuela was a coffee-exporting country. Coffee made up 75 percent of all its exports.[6] It was also a bankrupt country. Coffee prices had been going down in the world markets, the public debt was over $50 million, and the very high export taxes applied to coffee and cocoa had allowed Brazilian and Colombian products to displace Venezuela in the international markets. Venezuela had to import almost everything, "from machinery to needles."[7]

An epidemic of smallpox, civil war, and drastically reduced external credit dramatically colored the scene in Venezuela at the end of the century. It was then, 1899, that Cipriano Castro and Juan Vicente Gómez started a revolution from the Andes which was surprisingly successful. In October 1899 they were already in Caracas, in control of the government. But the economic life of the country was totally paralyzed. By 1902 the interest due on the external debt of Venezuela, $10 million, was higher than the national treasury could afford to pay.[8] In addition European countries were asking great amounts of money as compensation for the losses of their subjects during the revolution. By the end of 1902 these countries imposed a naval blockade on Venezuela's main ports, which was lifted only after a very weakened Venezuela accepted a forced settlement.

Oil would come to the rescue. In 1908 Juan Vicente Gómez replaced the ailing Castro as strongman. The Venezuelan economy was still in chaos, and Venezuela "had no diplomatic relations with Holland, France, Colombia and the U.S. and relations with Great Britain and Italy were very tense."[9] Gómez liberated all political prisoners, eliminated the import taxes on coffee and cocoa, and in April 1909 established a mining bureau.

The Shell group and Rockefeller's Standard Oil had by now developed an intense rivalry. Standard Oil tended to concentrate in the U.S. oil domestic market, whereas Shell had become very prominent in the international scene. In 1913 Shell entered the American market by buying a controlling interest in Roxana Petroleum Company, together with local investors. Deterding, head of Shell, claimed that "it was irritating to see a company, in any country, which operated . . . without the full cooperation of the nationals."[10] This was interesting since the issue of local participation would become extremely important in later years. By 1917 Shell was producing some 50,000 barrels per day more than Standard Oil, although the American company had been able to progress much more rapidly technically.[11] There is little doubt that it was this difference in early operating grounds—Standard Oil in the United States and Shell in the international trading scene—which was soon to account for some of the differences in organization which developed between these two giants. Standard Oil seemed to be much more task-oriented, more disciplined. Shell Oil companies tended to be more decentralized, with a stronger international component in their staff, which favored a more open, participatory decision-making. In turn these cultural differences helped to explain the relative degrees of adaptation and success of the two companies in their multiple international ventures. In Venezuela Shell's image was that of a benevolent grandfather, in contrast to the sterner, humorless father image projected by Standard. This perception would have significant economic importance: in the 1960s, during a period of active guerrilla warfare in Venezuela, Exxon's installations were considerably damaged while Shell's property remained virtually untouched. It was said, only half in jest, that Shell posted signs that read: "Creole's (Exxon's) pipeline is the other one." The interaction between the multinationals and the Venezuelan government, the political sector, and the public at large, has been strongly influenced by the way the different companies were perceived.

In December 1909 Gómez decided to grant a concession to explore, produce, and refine oil to the representatives of the English-owned Venezuelan Development Company, J.A. Tregelles and N.G. Burch.[12] This concession extended over 54 million acres, but would last only one year after the exploration period ended in December 1910. In 1911 and 1912 Gómez granted oil concessions to several of his friends. Covering millions of acres, these concessions went to A. Vigas, A. Aranguren, F. Jiménez, B. Planas, and, especially, R. Valladares. Valladares received concessions in twelve states, totaling 61 million acres.

Of course all these men did was to pass the concessions on to foreign companies which could develop them. Valladares negotiated with two American companies: General Asphalt and the Caribbean Petroleum Company.

In 1912 the Caribbean Petroleum Company hired geologist Ralph Arnold to conduct an evaluation of the extensive areas under concession. In slightly over a year Arnold and a group of Swiss, American, Dutch, English, French, and Venezuelan geologists and surveyors outlined nine promising areas. Arnold's saga is recorded in a fascinating account of those truly heroic early days of oil exploration in the country.[13]

Presenting his final report to Caribbean Petroleum, Arnold outlined some areas worthy of immediate attention, especially the area of Mene Grande, southeast of Lake Maracaibo. The company had run out of money, however, and was obliged to sell its interests. After looking in vain for an interested buyer in the United States, they were offered to Shell, who after much hesitation bought 51 percent, controlling stock for $10 million. Deterding would claim some years later that the riskiest gamble of his life was the acquisition of those rights, and would state that Arnold's report had been the basis for his decision.

In some of the areas outlined in Arnold's report—Quiriquire, Los Barrosos, Mene Grande, Cantaura, and El Mene—important oil fields would eventually be found. But Arnold also recommended areas such as Araya, Cumaná, Cubagua, and Coro where no oil has ever been found.

The members of Arnold's team, Wallace Gordon, Charles Eckes, Martín Tovar Lange, Luis Pacheco, and Rafaél Torres explored the area near the Guanoco asphalt lake in Eastern Venezuela in 1913, recommending the drilling of well Bababui-1, the discovery well of the Guanoco oil field, the first Venezuelan oil field after the La Petrolia modest discovery. The next year, the well Zumaque-1 was drilled in the Mene Grande area, southeast of Lake Maracaibo. It became the discovery well of the Maracaibo Basin. The second well of the Mene Grande field, Zumaya-1, took 10 months to reach final depth at some 1,500 feet but had an initial production of about 40,000 barrels per day. This lasted only a few days until it became stable at 2,000 barrels per day, but it was enough to greatly excite Shell's interest.

The First World War retarded the development of the newly discovered oil fields until activity was reinitiated in 1917. Another company controlled by Shell, Venezuela Oil Concessions, discovered the La Rosa field, on the eastern shores of Lake Maracaibo, with the Santa Barbara-1 well. This field became famous in 1922, when the well Los Barrosos-2 was completed and blew out, producing at a rate estimated at 100,000 barrels per day. This caused a rush into the country rather like the California gold rush of the previous century. By now Shell had a huge lead over other competitors, since it was already refining and exploiting hydrocarbons from the area east of Lake Maracaibo and had made important oil discoveries in the Maracaibo Basin, both east and west of Lake Maracaibo.

In 1926 Gulf Oil discovered the Lagunillas field with the well Lago-1. This discovery rapidly led to other important finds in the surrounding areas of Tia Juana and Bachaquero. In fact what had been discovered was an almost continuous string of oil fields east of Lake Maracaibo which in time received the generic name Bolívar coastal fields (BCF), since the fields lay in the District of Bolívar, State of Zulia. The BCF became one of the three largest oil fields in the world.

Not all efforts were equally successful. Most were not successful at all. Companies such as Cities Services, Union Oil of California, Compagnie Française de Petrole, Compañia Española de Petróleos, and the New England Oil Company all lost both time and money in Venezuela. More fortunate were the bigger multinationals such as Shell, Exxon, Gulf, Mobil, and Texaco, with more lasting power, more money to risk. Also winners were the friends of Gómez, the Venezuelan dictator. As early as 1923 Gómez and a group of Venezuelan friends, Lucio Baldó, Roberto Ramírez, G. Gonzalez Rincones, and a Mr. E.B. Hopkins, had formed the Compañia Venezolana del Petróleo, which recovered holdings in several states of Venezuela.[14] The company transferred most of these holdings to another company, called South American Oil and Development Corporation, which in turn signed an agreement with José Gil Fortoul, on the basis of which the well-known Venezuelan lawyer, writer, and Gómez go-between agreed to secure this company with holdings from the Compañia Venezolana del Petróleo, in return for 40 percent of the net profits. This percentage was no doubt split with Gómez and the rest of the group. These practices, combined with the earlier practice of assigning concessions to Venezuelan individuals who in turn sold them to foreign corporations, made the instant fortune of many Venezuelan families, who are now totally accepted socially. This brings up an interesting moral issue, since the criticism of the political sectors of the country has always been reserved for the multinational corporations whereas corrupt Venezuelans who made immense fortunes at the expense of their countrymen have almost never been criticized.

Early Hydrocarbons Legislation

Between 1917 and 1935 the Venezuelan government issued six different versions of the hydrocarbons law. The continuous process of revision was largely the result of the work of one man, a medical doctor by the name of Gumersindo Torres, who served as Gómez's minister of development from 1917 to 1922 and, again, from 1929 to 1931. Torres separated the hydrocarbons from other minerals for the purpose of oil's legal regulation. He advocated a moderate, middle-of-the-road approach to law-making. He would say, "Laws should not be too severe nor too complacent. The Ministry of Development must pursue a moderate course."[15]

The hydrocarbons law of 1920 was the first comprehensive legal document specifically applied to hydrocarbons. It contained several unrealistic clauses, however, like article 50,   obliging concessionaires to exploit their holdings "within three years." Article 8 reserved to the owner of the land the right to ask for an exploration permit, but this was also unrealistic as these Venezuelan landowners could only hope to transfer their rights to the foreign companies. As a consequence the original law was rapidly modified. Minister Torres issued the 1921 hydrocarbons law and one year later still another law, which with minor changes remained basically intact until 1943.

Perhaps the main contribution of Gumersindo Torres to the issue of control of the Venezuelan oil industry by the state was the creation of the Technical Hydrocarbons Service. Torres said, in the memoirs of the Ministry of Development for 1930, that such an organism had been created "to monitor both the technical aspects of the exploitation and the companies themselves." In order to train the required technical staff, Torres sent a group of six young engineers to Oklahoma to study petroleum engineering. Three (Siro Vasquez, Manuel Guadalajara, and Jorge Hernandez) went to the University of Tulsa and the other three (Edmundo Luongo, Abel Monsalve, and Jose Antonio Delgado) went to the University of Oklahoma. Vasquez would have a very distinguished career, reaching the senior vice presidency of Exxon. Edmundo Luongo would become minister of Mines and Hydrocarbons. All of these men eventually became a credit to the Venezuelan civil service and established very high standards of honesty and excellence which were maintained for many years. The University of Tulsa became a major training center for Venezuelan students in the field of petroleum. Today no less than 500 Tulsa alumni are Venezuelans, an impressive number considering that the university has always been rather small.

According to W.M. Sullivan, Venezuela "structured the most liberal petroleum policy in Latin America.[16] To attract foreign investments, exploration and exploitation tariffs were very low and the same applied to royalties. . . . Import duties for petroleum equipment were waived." Gómez favored "the presence of North Americans in our country because where they are, there is money. . . . They are hard workers and . . . have never meddled in our internal affairs. There is no reason to be afraid of them. We will take all appropriate measures to protect our rights."[17] With this enthusiastic endorsement coming from the man who ruled the country with an iron hand from 1908 to 1935, it was easy to explain why U.S. petroleum investments in Venezuela flourished.[18] Although coffee remained the main Venezuelan export until about 1920, oil rapidly became more dominant in the Venezuelan economy. In 1917 production was 121,000 barrels. By 1921 it had increased to 1 million barrels, and by 1930 it had reached 135 million barrels. As the oil industry grew by leaps and bounds, the country was transformed, though not always positively. Activity concentrated along the eastern shores of Lake Maracaibo as well as in Eastern Venezuela, in the states of Anzoategui and Monagas. Lake Maracaibo started to suffer the effects of industrial pollution. Rural exodus became significant. The town of Maracaibo grew from 40,000 people in 1900 to 120,000 in 1929. House rentals increased by 900 percent.[19] Thousands of skilled laborers from Trinidad flooded into the area, usually earning higher salaries than their Venezuelan counterparts, eliciting a xenophobia until then largely unknown. Shantytowns were created overnight next to the oil field camps. Prostitution, drinking, and gambling became familiar aspects of a scene so recently bucolic.

The Emergence of Popular Resentment

It was at this point in the relationship between the oil companies and the country that a reaction against the foreign "invaders" became measurable. As is usually the case the leading forces behind this reaction were the intellectual minorities. As early as 1912 a novel had been published in which the villain was called John Smith, a common American name.[20] In the story an asphalt mine falls under the rapacious control of Smith and of Morgan, (another suggestive name).

Novelist Jose Rafael Pocaterra described the oilmen as "the new Spaniards." He wrote in 1918,

> One day some "Spaniards" mounted a dark apparatus on three legs, a grotesque stork with crystal eyes. They drew something (on a piece of paper) and opened their way through the forest.[21]

and added,

> Other new Spaniards would open roads . . . would drill the earth from the top of fantastic towers, producing the fetid fluid . . . the liquid gold converted into petroleum.

Rufino Blanco Fombona, a Venezuelan writer and politician who spent long years in exile during the Gómez regime, published in 1927 the novel *La Bella y la Fiera*, in which he described the conflict between the Venezuelan worker and the foreign boss:

> The workers asked for a miserable salary increase and those blond, blue-eyed men who own millions of dollars, pounds and gulden in European and U.S. banks, refused.

This perception of the foreign oil-company representatives as heartless new conquerors became deeply rooted in the social conscience of the Venezuelan people and never really disappeared. In many ways the reaction against foreign capital led by the novelists of the period 1920-1940 can be considered the first real step toward the nationalization which would take place not until 50 years later. Even today feeling runs high against the presence of foreign companies in Venezuela. The strong sentiment is based in the historically frequent association of these enterprises with the dictatorial governments which dominated the country for many years. This resentment has never been entirely understood by foreign corporation executives, who always believed that they were doing legitimate business in the country and that they could not have been expected to become involved in local politics. This is clearly a difficult gap to bridge since Venezuelans tended to see in this neutrality a subtle variety of political involvement. By not being against dictatorships foreign corporations appeared, in fact, to be for them.

The basic sociological changes in Venezuela resulting from the development of the oil industry have been summarized by Gustavo Luis Carrera. The primary changes were the following:

Disruption of traditional Venezuelan values and ways of living

Complicity between a corrupt Venezuelan elite and the foreign corporations to exploit the oil to their advantage

Racial discrimination

Obsession with attaining instant wealth

Rural exodus

Emergence of crime and vice

Deterioration of the quality of life in the oil-producing areas

Takeover of lands and forests for oil exploitation[22]

Carrera's analysis is important although it centers, mostly, on the negative changes brought about by the oil industry. Positive changes should also be listed:

Economic prosperity

Creation of a basic infrastructure of roads, schools, and hospitals in the rural areas

Installation of organized communities—oil camps—in remote areas, which created a new awareness among Venezuelans of how a modern, more comfortable life could be led

Emergence of an increasingly important middle class

Emergence of an organized labor movement.

By 1935, at the time of the death of dictator Juan Vicente Gómez, Venezuela was a full-fledged oil-producing country. Production was slightly over 400,000 barrels per day. Oil contributed 25 percent of total government income (Bs 60 million out of a total Bs 210 million). Still, the economy and general social conditions of the country were extremely poor. An American mission visiting the country in 1940 had this to say:

> The capacity to read and write of the great majority of the population is low. . . . The standards of living and buying power of the people are very low. In many Venezuelan cities there are no sanitary facilities, which affects negatively the health of the population. . . . Food supplies are insufficient and of low quality.[23]

Venezuela was no longer a rural country. A new political generation had come to life with the death of Gómez. This group started to denounce publicly the dramatic social conditions existing in the country. In March 1936 the Caracas bullring was the scene of "the first political mass event ever held in Venezuela in the twentieth century."[24] At this event Betancourt said,

> (Venezuela), a country without external debt, true, but in the hands of the most audacious and aggressive sector of international finances, the oil sector. . . . This is a country in which 80 percent of exports are made up of petroleum, an industry which is in foreign hands.[25]

The Years of Lopez Contreras

Gómez's successor was Eleázar López Contreras, a man who had accompanied him since the first day of his political career. His years in power, from 1936 to 1941, are still very much the subject of debate between those who would have wanted a rapid and profound process of democratization and those who preferred a cautious transition from three decades of brutal dictatorship to parliamentary democracy. During the initial months of the López Contreras tenure, the oil workers went on strike, asking for a basic minimum daily salary of Bs 10 ($3), cool and potable water in the working areas, Sundays off, and a few other things. The oil companies refused, and a strike was called. This strike was not an orthodox affair. It had all the characteristics of a crusade, with intellectuals, businessmen, and students all joining in support of the striking workers.[26] Government labor inspectors sent to analyze the reasons for the strike and to recommend possible solutions openly sided with the workers. The main government envoy, Dr. Gutierrez Alfaro said that while the unions had been conciliatory, the company representatives had proven very reluctant to make any concessions. After 43 days, the government ordered strikers to return to work. A salary increase of Bs 1 ($.030) per day was decreed, and the companies were instructed to supply cool, potable water to the work sites. Clarence Horn, writing for *Fortune* in March 1939, described the living conditions of the average Venezuelan oil worker in the following fashion:

> He lives in a barrack built on stilts on the Lake of Maracaibo. . . . To bathe he can use lake water but will have to skim off the layer of oil. He will have to buy the water that he drinks at home. His diet consists of corn, bananas, black beans, rice, for which he has to pay 5–6 bolivares per day. . . . At the end of the month, he will have no money left but surely will have a venereal disease which the company doctor will cure for free.[27]

The López Contreras government had positive achievements. The Central Bank of Venezuela was created, a very advanced work law was passed, and local elections were held in which the opposition to the government obtained important victories.

Oil concessions granted during this period required concessionnaires to refine locally a portion of the crude oil produced. This stipulation led in 1939 to the installation, in Eastern Venezuela, of the Caripito refinery, with capacity of 30,000 barrels per day. Fiscal income from oil in 1938 was about Bs 115 million, but import exemptions totaled about Bs 100 million. This left only Bs 15 million as net government income in exchange for about 150 million barrels of high-quality oil which were gone forever. Obviously this would not do.

In January 1936 the well La Canoa-1 was completed. This was the discovery well for the immense heavy-oil deposits of the Orinoco Belt. At the time, however, this heavy and impure oil was not commercially attractive. Forty years would go by before the area was put under systematic development. In 1937 the local subsidiary of Gulf drilled and completed the well Oficina-1, the discovery well for the Oficina field of eastern Venezuela.[28]

In 1938 the government passed a new hydrocarbons law. Earlier in the year, Mexican President Lázaro Cárdenas had decreed the takeover of the foreign oil company assets by the government, an action which caused a wave of nationalism in Latin America. The new Venezuelan law was important for several reasons. It gave the government explicit authority to create national oil companies, although this option was not exercised until 1960, when the first Venezuelan state oil corporation was created. It also reduced drastically the tax relief on imports traditionally given to oil companies and increased exploration and exploitation taxes.

Without doubt, under the presidency of López Contreras, the Venezuelan government significantly increased its share of administrative control over the oil industry. Financial participation also increased, although much less substantially. The new political climate in the country was favorable for this evolution since, for the first time in the century, there was political freedom and social ferment. By 1938 the oil industry was clearly the economic lifeline of the country. Venezuela had become the second largest oil producer in the world. It was no longer a country dependent on coffee. It had become a country dependent on oil.

Main Characteristics of the Period

Technical Advances

The 25 years between 1915 and 1940 constituted a period of consolidation for the Venezuelan oil industry. After the discovery of the prolific Maracaibo Basin, the attention paid to Venezuela by foreign companies as a potential oil country was great. These companies applied new technological tools to the discovery of hydrocarbons. Venezuela became the testing ground for new prospecting methods. Although a majority of the discoveries between 1915 and 1930 were still due to the direct observation of oil seepages by field geologists, much of the success was due to the use of new techniques. As early as 1920 Shell geologists extensively used hand-drilled auger holes to obtain fresh rock samples from near-surface but unexposed bedrock. In 1920 as well Dr. Carlota Maury, a paleontologist, was engaged by Shell to study the fossil molluscs of the Guanipa area, in eastern Venezuela.[29] On the basis of this work, she could advise Shell on where to drill for best results. In 1925 the torsion balance was introduced in Venezuela to detect gravity changes in the subsurface and, therefore, to predict the nature of the underlying rocks. In the late 1920s the seismic reflection method started to be used in the prospecting for oil in Venezuela. The first commercial electric log was run by Schlumberger in 1929, in well R-216 of the La Rosa field, east of Lake Maracaibo.

Slow Progress in Government Control of the Industry

The degree of control of the oil industry on the part of the government remained low during this period. This was a product of the existing political climate and the lack of Venezuelan know-how. Dictator Gómez seemed to trust foreigners more than he did his fellow countrymen. He saw oilmen as hard-working, knowledgeable, interested in making money, and, above all, disinterested in local politics. Oilmen saw Gómez as a shrewd, greedy, and strong politician who should be kept happy if they were going to keep doing business in Venezuela. They could not or did not want to see too far into the future and decided to make the most of the present. For many years this relationship was profitable for both parties but, when it ended, the foreign companies started to pay a high price for having had it.

At the same time there were no Venezuelans who had the knowledge to exercise control over the industry. We have already seen that the first Venezuelan engineers trained in petroleum came out of school only in 1935. The first organized attempt at government supervision of oil operations came in 1930 with the creation of the Technical Hydrocarbons Service. The intention was already there but more than 40 years would pass before the government could say it had real control over the industry.

Increasing Nationalism

The close relationship between the foreign oil companies and the Gómez dictatorship created a deep popular resentment against oilmen. Although it was true that the foreign oilmen could not afford to take the side of the people against the dictatorship, it was also true that, in general, they were perceived as taking the opposite side, that of the dictatorship against the people. Many of the then-young opposition political leaders would harbor all their lives a deep mistrust of the motivations of foreign oil companies, even after the old oilmen had been replaced by younger, more liberal and more sensitive executives. Even a moderate democrat and good friend of the United States like Rómulo Betancourt had this to say about the relations between the foreign oil companies and the corrupt Gómez regime:

> The friendship between Gómez and Queen Wilhelmina was well known. . . . The Royal Dutch swam in an ocean of concessions. Holland captured a boat manned by Gómez's adversaries. . . . No less effective were the cooperation and support of the governments of his British Royal Highness and of the United States. The intelligence services of both countries kept the dictator informed of any potential risks (that could endanger his government).[30]

It is possible that this adverse reaction against foreign oil companies and oilmen was, after nationalization, transformed into a distrust of the Venezuelan oil industry and of the Venezuelan oil executives who were perceived as the heirs of the original Gómez supporters. It is ironic that with few exceptions the Venezuelan oil executives were men of the middle class whose families had had no link with the Gómez regime, whereas some of the politicians who attacked them were members of families with a history of connections and dealings with Gómez and his friends.[31]

Notes

1. "Petroleum Developments in Venezuela during 1930," *American Institute Mining and Metallurgical Engineers Transactions* 42 (1930), pp. 522–531.

2. The English version of this work is *Oil, the Making of a New Economic Order: Venezuelan Oil and OPEC* (New York: McGraw-Hill, 1975). The quotations in the present book have been translated by me from the original.

3. Anibal Martínez, *Cronología del Petróleo Venezolano* (Caracas: Edic. Foninves, 1976). First published in English as *Chronology of Venezuelan Oil* (London: Allen and Unwin, 1969).

4. Ibid., pp. 27–34.

5. Vallenilla, *Auge, Declinación y Porvenir del Petróleo Venezolano* (Caracas, 1973), p. 19.

6. William M. Sullivan, "The Rise of Despotism in Venezuela: Cipriano Castro, 1899–1908," Unpublished PhD diss., University of New Mexico, Albuquerque, 1974, p. 26.

7. Ramón Veloz, *Economía y Finanzas de Venezuela, 1830–1944* (Caracas, 1945).

8. Fundación John Boulton, *Política y Economía en Venezuela, 1810–1976* (Caracas, 1976), p. 237.

9. Ibid., p. 249.

10. Vallenilla, *Petróleo en Venezuela,* p. 13.

11. Raymond Vernon, Harvard Institute for International Development, Development Paper 3, Harvard University.

12. Martínez, *Cronología del Petróleo Venezolano,* p. 54.

13. Ralph Arnold, *The First Big Oil Hunt* (New York: Vantage Press, 1960).

14. In only two years this company obtained profits of more than $8 million, simply by transferring concession titles to foreign companies. Some of the Venezuelans connected with transfers were Hernan Rodríguez, R. Hernández Ron, Octavio Pérez, Julio Lira, José Tovar, José M. Cárdenas, Juan Pietri, and José Rengifo. See Vallenilla, *Petróleo en Venezuela,* p. 56.

15. Vallenilla, *Petróleo en Venezuela* p. 82.

16. W.M. Sullivan, "Situación Economica y Política durante el Período de Juan Vicente Gómez," in *Política y Economía en Venezuela, 1810–1976* (Caracas, 1976), p. 258.

17. Ibid., p. 259.

18. U.S. investments in Venezuela grew from $11 million in 1924 to $160 million in 1928.

19. Sullivan, "Situación durante el Periódo de Gómez," p. 259.

20. Daniel Rojas, "Elvia," analyzed by Gustavo L. Carrera in his excellent book, *La Novela del Petróleo en Venezuela* (Caracas, 1972).

21. Carrera, ibid., p. 12.

22. Ibid., p. 80 ff.

23. Quoted by J.A. Mayobre in *Política y Economía en Venezuela, 1810-1976* (Caracas, 1976), pp. 275-76.

24. See R. Betancourt, *Venezuela, Política y Petróleo,* Seix. Barral (Caracas, 1978).

25. Ibid., p. 103.

26. Ibid., p. 111.

27. Quoted by R. Betancourt, ibid., p. 112. Courtesy of "Fortune" magazine; © 1939 Time Inc.

28. By 1980 the Oficina field had produced over 700 million barrels of mostly high-quality oils.

29. R.J. Forbes and D.R. O'Beirne, *The Technical Development of the Royal Dutch Shell, 1890-1940* (Leiden, 1957).

30. Betancourt, *Venezuela, Política y Petróleo,* pp. 95-97.

31. Many Venezuelan fortunes were made during the days of Gómez by landowners who collected royalties and other benefits from the oil companies.

2

The Tortuous Road to Democracy, 1941-1970

The Medina Years

In 1941 President López Contreras handed over the government to his chosen successor, Isais Medina Angarita, also an army general from the Venezuelan Andes, a member of the elite which had been ruling the country since the turn of the century. Congress was dominated by López Contreras, and the naming of Medina Angarita was never in doubt. However, the thirteen members of the opposition deposited their votes for novelist Rómulo Gallegos, in what was a symbolic protest since Gallegos was not an official candidate.[1]

Medina Angarita proved to be a true democrat. He legalized the Partido Democrático Nacional (PDN), which later became Acción Democrática (AD). During his government the first income tax and agrarian reform laws were passed. Even more important was the enactment, in 1943, of a new hydrocarbons law. This law essentially converted the old concession titles into new titles valid for forty more years but subject to new conditions more favorable to the government.[2] Taxes were increased, the principle of reversion of assets reinforced, and the concept of a 50/50 split of the profits between government and the oil industry was introduced.[3] This split did not materialize until 1948, however.

The new hydrocarbons law proved controversial. Independent members of the National Congress and the small but very vocal Acción Democrática group led by Dr. Juan Pablo Pérez Alfonzo strongly argued against portions of the law. Criticism from this group centered on the following points:

1. The mechanism to select exploitation parcels seemed to be too advantageous to concessionaires.
2. Old concessions were renewed for forty more years.
3. All pending claims against the concessionaires were dropped.
4. The 50/50 profit split was still not good enough.
5. The law was the product of negotiation and was thus perceived as a settlement rather than as a unilateral, sovereign decision.

19

The supporters of the new law included some of the most respected lawyers and oil experts in the country. They claimed that the law unified all previous concessions under a single, coherent, legal instrument and that it provided the nation with a much larger participation in the profits as well as with greater administrative control.

The foreign oil companies were not very happy about the law. Some of the industry's top executives were very reluctant to accept it. The negotiations have been described as follows:

> When the Venezuelan government was unable to make any progress with oil company officers in Venezuela, it decided to turn to the United States' government. Discussions between the Attorney General of Venezuela, Gustavo Manrique Pacanins, top officers of the oil companies and the U.S. Department of State went on for months. Attorney General Manrique finally took the position that unless a satisfactory settlement could be reached by friendly negotiations, the Venezuelan government would be forced to seek its objectives by unilateral action.[4]

Geologist Wallace Praat, a top executive of Standard Oil of New Jersey, became the leader of the group favoring a settlement with the Venezuelan government.

> He held that the issue was a question of equity because the company was making higher profits than had been expected. He also held that foreign operations in the future would survive not on the strength of contracts but on the basis of recognition by both sides of a mutuality of interests and on an equitable sharing of benefits.[5]

Negotiations finally bore fruit. In reaching a satisfactory agreement, the Venezuelan government was assisted by Herbert Hoover, Jr., and A.A. Curtice. A letter written by Wallace Praat in 1964 illustrates the value of tact when dealing with sovereign governments:

> In 1942 the situation which confronted (Standard Oil) Jersey and other foreign companies in Venezuela became, in my judgment, almost exactly a duplicate of the situation we had faced in Mexico in 1937. Again a Jersey Board insisted that we should not renegotiate our concessions. Again I took the opposite view and proposed that we proceed to renegotiate . . . this time I was successful. Eventually the Board reversed itself and authorized me to go to Venezuela and handle renegotiations so far as Jersey was concerned. I did so and the other companies followed Jersey's lead. The result was that Jersey and the other foreign companies came out of renegotiation proceedings in a far more advantageous position than they were in when the renegotiations started.[6]

The law, once passed, remained basically unchanged until 1975, the year of nationalization, with only two partial revisions being made in 1955 and

1967. The combined effect of this law and of the income tax law passed in 1942 allowed Medina's government a significantly greater financial participation in the profits of the oil industry. In 1944 the government oil income more than doubled as compared to 1943.

The government also granted new concessions in 1944. World oil demand was increasing rapidly due to the war. By 1945 Venezuelan oil production already averaged close to 1 million barrels per day. The new concessions were being explored and manpower increased substantially to about 45,000 employees. Important discoveries of light-gravity oils were made in the Maracaibo Basin, especially west of Lake Maracaibo. The country was in an oil boom.

Although the general economic situation of the country was better than it had been for many years, the political atmosphere was far from good. Acción Democrática accused the government of corruption. In March 1944 the National Labor Convention had been dissolved by the government on the grounds that the labor movement had become politicized. The situation seemed to improve when Acción Democrática agreed with the government on the name of Diogenes Escalante, Venezuelan ambassador to the United States, as the replacement in 1946 of President Medina. The sudden mental breakdown of Escalante eliminated that possibility, however, and Medina went on to choose a little-known politician, Angel Biaggini, as his replacement.

Acción Democrática Takes Over

The adverse reaction of Acción Democrática to this choice did not change Medina's mind. As a result the party took the view that all avenues for political evolution were closed and that only a revolution could force a democratic change. In October 18, 1945, the Medina government was overthrown in a coup led by the party and a group of army officers.

The new government wasted no time in increasing oil company taxes. In December 1945 a decree was issued raising corporate taxes to all companies with earnings above Bs 2 million. Of the new amounts collected, at least 98 percent came from the oil companies. Clearly this was a law "focused on the oil companies, as Betancourt admits.[7] The oil policy of Acción Democrática became the main economic guideline of the government for the next three years, from the moment Rómulo Betancourt became president of the Revolutionary Junta in October 1945 until December 1948 when President Rómulo Gallegos was overthrown by a military coup. This policy, as stated by Betancourt, had the following ingredients:[8]

Tax increases within reasonable limits.

Direct sale of oil by Venezuela in the international markets, using the "royalty" oil. (Venezuela had the option of receiving up to 16.66 percent of the production in kind as royalty.)

No more concessions to private enterprises. The creation of a state oil company to develop national reserves.

Processing the oil domestically and building a national refinery.

Applying strict conservation principles in the production of oil and utilizing the natural gas, which was largely flared.

Requiring foreign companies to reinvest a part of the oil profits in the development of the agricultural and livestock sectors.

Substantial improvement of salaries and living and working conditions of the Venezuelan oil workers.

Utilization of a substantial share of the oil income obtained through the new tax schemes in the creation of a fully Venezuelan, diversified economy.

The government made clear from the start that the nationalization of the oil industry was not being considered. To explain this moderate attitude, Betancourt emphasized the differences between Mexico and Venezuela. Mexico had, he claimed, a more diversified economy than Venezuela, where 92 percent of all imports were financed by oil revenues. Betancourt wrote many years later that nationalizing the oil industry at that time would have been suicidal.[9] Still, after many years of an essentially happy marriage with Gómez and of relative comfort under López Contreras and Medina, the oil companies viewed the new government with distrust. Betancourt did not hide similar feelings. It is interesting to see how the best-informed and shrewdest of Venezuelan political leaders have always had problems distinguishing between what is essential and what is only anecdotal in American–Venezuelan relations. In his book on oil, for example, Betancourt would quote a college professor:

> Professor Lester C. Uren, from the University of California and author . . . of a book about oil . . . said: "The (Venezuelan) government was overthrown in 1945 and the new government intends to raise taxes. Their socialistic tendencies have created less favorable perspectives for foreign capital."

and worry:

> The suggestions of retaliation implicit (in Professor Uren's opinions) were not strong enough to stop us from going ahead with our commitments. . . . We dictated a decree stipulating a higher tax.[10]

What clearly appeared to be no more than a very personal opinion from a member of academia seems to have been taken by Betancourt as an expression of American policy. Such cases of hypersensitive reaction from Venezuelan political leaders to isolated external criticism have always been abundant and have distracted considerable talent and effort from more fruitful endeavors.

In 1946 Venezuelan oil production averaged slightly above 1 million barrels per day. Already 54 percent of the budget was financed by oil revenues. This year was the first in a cycle of rapidly increasing national dependence on oil.

In December 1947 Acción Democrática candidate Rómulo Gallegos was elected president by a great majority. This was the first time in Venezuelan history that a president had been directly elected by the people by universal, secret balloting. Gallegos was a respected novelist with a pure, perhaps naive, attitude toward politics and with a great sense of personal integrity. He would have made a wonderful president of Switzerland. In the politically agitated climate of the 1940s in Venezuela, he was a misfit. Nine months after his inauguration he was overthrown by a military coup led by essentially the same officers who had played a role in bringing his party to power two years earlier.

A New Dictatorship

In his last message to the country, Gallegos blamed the multinational oil companies for his ousting when he mentioned the "powerful economic forces, those of local capital without social sensitivity and, perhaps, also those of the foreign exploiters of our subsurface (mineral) riches. . . . They have been, I do not hesitate in denouncing them, the ones who have encouraged the desire for power of (the) military officers."[11] Gallegos felt strongly that the coup had been largely in retaliation against the tax increases imposed on the oil companies. These increases had taken place only 12 days before the coup, on November 12, 1948, and had been based on a modification of the income tax law. It was at this moment that the Venezuelan government finally came to obtain about half of the oil profits. The finance minister of Gallegos, Manuel Pérez Guerrero, described the objectives of the measure thus: "In this fashion the law will stipulate that the participation of the Nation will never be lower than that of the companies."[12] Although it is true that the Acción Democrática government increased taxes, established strong ties with Venezuelan labor by creating both the Venezuelan Confederation of Labor and the Petroleum Workers Federation (Fedepetrol), and tried to enter directly into the world oil markets, it is doubtful that the oil companies were, as Gallegos claimed, behind the coup that ousted him. This coup seemed to be the result of a dominantly internal struggle between the army and a populist and aggressive political party which probably tried to do too much, too soon. The strong popular support that Acción Democrática seemed to have did not materialize when it was needed and Gallegos was practically abandoned by the people who had sworn to defend him and his government. In April 1947 the Petroleum Workers Federation had issued a strong warning:

> The position of the petroleum workers . . . is very definite. We resolutely back the National Constituent Assembly and the government emanating from it against whatever insurrectionist attempt or simple conspiratory provocation. We are disposed to respond with the means that we have, we will abandon the petroleum installations in all Venezuela in order to (fight) against the enemies of democracy and the working class.[13]

However, they did not fire one single shot when the time came.

The military junta which took over dissolved the workers' unions and drastically curtailed political activity in the country. Meanwhile, after a temporary lull in the U.S. oil demand due to a mild winter season in 1949–50, demand for oil had kept increasing. The invasion of South Korea by North Korea and tension in the Middle East had led to a greater dependence of the United States on Venezuelan imports. Venezuelan oil exports increased rapidly, from 460 million barrels in 1949 to about 600 million barrels in 1951. Wells drilled jumped from 625 in 1949 to almost 1,200 in 1951. Venezuela was in another oil boom. Political ambitions grew at the same rate. In November 1950 the head of the military junta, Lieutenant Colonel Carlos Delgado Chalbaud, was murdered by a group of paid assassins and the number-two man, Lieutenant Colonel Marcos Pérez Jiménez, came into power.

The dictatorship of Marcos Pérez Jiménez lasted until 1958. The perception of the public was that relations between the U.S. government and oil companies and the Venezuelan dictator were much more cordial than they had been during the democratic government of Acción Democrática. Venezuela started to receive massive U.S. military aid. "Diplomatically, militarily, and economically, Venezuela remained firmly within the sphere of influence of the United States."[14]

In 1950 the military government created the Ministry of Mines and Hydrocarbons to handle all matters pertaining to the exploitation of hydrocarbons and minerals. Because of the favorable climate for private capital, foreign investments in Venezuela more than doubled during the 1950s, reaching $2.5 billion in 1960. Of this amount nearly $2 billion were in oil. "U.S. exports to Venezuela surpassed one billion dollars in 1957, making Venezuela, a nation of only seven million people, the sixth best market in the world for U.S. traders."[15] The country became extremely dependent on oil and even the most basic commodities were imported: Colombian meat, black beans from the Dominican Republic, frozen vegetables from the United States. Whereas government income rose in absolute terms, the participation of the state expressed as a percentage of the earnings before taxes actually decreased, going from about 55 percent in the years 1948-1952 to about 52 percent for the period 1954-1957. This percentage would suddenly increase to 65 percent in 1958, after the overthrow of the military dictatorship.

In 1956-57 the Pérez Jiménez government granted new oil concessions. About 1.6 million acres of new territory were granted to about fourteen foreign oil companies. The government invoked mainly economic considerations for this decision. This was hard to justify as government oil income in those years had been considerable. However, the Pérez Jiménez government had engaged in grandiose building projects requiring even more money. The granting of these new concessions gave the government about $600 million in bonus payments.

The granting of these concessions produced an abrupt but temporary increase in exploration and drilling activities. From 1,200 wells drilled in 1955 the figure increased to more than 1,700 in 1957. In general, however, the impact of the new concessions on the general level of activity of the industry was short-lived. By 1960 the companies had returned to the nation about 70 percent of the areas obtained in 1956-57. The areas kept were mostly in Lake Maracaibo. The net effect of the granting of new concessions was to increase even further the Venezuelan dependency on oil.

At the same time this was happening in Venezuela, the United States—its principal client—was starting to develop great concern over its increasing levels of crude oil imports, most of which were coming from Venezuela. From 1950 to 1960 U.S. oil imports had increased from 174 million barrels to about 400 million barrels. In 1954 a cabinet committee on energy supplies and resources policy was formed to analyze the relationship of oil imports to national security. Although President Eisenhower was personally opposed to petroleum quotas, the committee "found that oil import control was an important element in national defense, specifically stating that should the ratio of imports to domestic production rise above 1954 levels, security problems would be created for the United States."[16] Congressional representatives from oil-producing states such as Texas and Oklahoma started to call for mandatory regulations.

In spite of the intense pressure, President Eisenhower agreed only to a voluntary program of imports control. Even then residual oils and Canadian and Venezuelan crude-oil imports were excluded from the voluntary restrictions. Ideas on imposing a more formal control were temporarily abandoned during the Suez Crisis of 1956, but in 1957 a new cabinet committee was formed which recommended a ceiling for 1957 U.S. oil imports.

On the basis of this recommendation, a system of import quotas came into being which would last in one form or another until 1973, when it was suspended as a response to the energy crisis.

The impact of mandatory oil-import quotas on the Venezuelan oil industry was very negative. The Venezuelan government felt that the country was being discriminated against. Mexico and Canada had been given special concessions, whereas Venezuela oil imports were subject to restrictions. As a result the level of Venezuelan crude-oil imports entering the United States decreased substantially, from 47 percent of the total volume in 1959 to about 20 percent of the total in 1970 (from 450,000 to 268,000 barrels per day).

Oil prices in the world market weakened noticeably in 1958 due to the increase in the production in the Middle East and Africa. British Petroleum lowered posted prices by about 15¢/barrel in Iraq, Kuwait, Iran, Qatar, and Saudi Arabia.

In January 1958 a coup took place in Venezuela and a civilian-military junta replaced the dictator. The new government decided in December of that year to increase taxes and to apply this increase retroactively to the earnings obtained during the year. This decision increased the government share of the oil profits from 52 percent to about 65 percent.

Betancourt and Pérez Alfonzo

The national elections of 1958 put Acción Democrática back in power. Rómulo Betancourt was elected president and Juan Pablo Pérez Alfonzo was again put in charge of conducting Venezuela's oil policy. This time Betancourt was seen with much greater sympathy by the U.S. government. His main rival, Wolfgang Larrazábal, had accepted the endorsement of the Communist party, and this worried the U.S. Department of State to the extent that their feared old foe Betancourt now became the better alternative.

The Betancourt government did have a clear oil policy. It centered on five principles:

No more oil concessions would be granted to foreign companies.

Oil prices would be defended at all costs.

Venezuela would continue looking for a hemispheric preferential treatment, especially from the United States.

A national oil company would be created.

Venezuela would promote the creation of an organization of petroleum-exporting countries:

These points became known as the "Pentagono Petrolero" of Pérez Alfonzo, the oil pentagon.

No More Oil Concessions

? Alf. believed that prices would rise again + therefore want to exhaust reserves

When Pérez Alfonzo announced the decision of the government not to grant additional concessions to foreign oil companies, one of the immediate results was the almost total cessation of exploration activities in the country. On the one hand the oil companies claimed that their holdings were already explored and that no important oil prospects remained to be evaluated in detail. On the other hand they refrained from exploring in free acreage once they knew they could not have eventual access to it. As a result, for more than fifteen years, 1958–1973, there was little or no exploration in Venezuela and consequent stagnation in the level of hydrocarbons reserves. Proven reserves fell from the 17 billion barrels of 1959 to 13.8 billion barrels by 1973. The volume of reserves in turn influenced the producing life expectancy of the country. In 1958 this life expectancy was of about 20 years, whereas in 1973 it had dropped to about 12 years. As a result of this unfavorable ratio and the increasingly old age of Venezuelan oil reservoirs, production output reached a peak of 3.7 million barrels per day in 1970 and had dropped to about 2.3 million barrels per day in 1975, the year of nationalization.

There is no doubt that the deterioration of the oil-producing capacity of Venezuela was partly due to the no-more-concessions policy. Since new fields were not found, the old ones had to be slightly overproduced and became shorter-lived.

A distinguished Venezuelan writer and publicist, Arturo Uslar Pietri, strongly criticized the no-more-concessions policy on the grounds that it had not been accompanied by a substitute policy. Uslar Pietri claimed that Dr. Pérez Alfonzo had closed a door without opening another. Public opinion was generally in favor of the decision since it suited their nationalistic preferences, but the more informed sectors worried about the long-term effect of the policy on the vitality of the oil industry. It was only in 1961 that a new concept—service contracts—was announced as a replacement to the no-more-concessions policy, but it would take ten more years for the first contracts to be negotiated. Exploration came to a virtual standstill.

At least one analyst of oil matters, Luis Vallenilla, claimed that Pérez Alfonzo actually wanted to see the oil industry decline and that "he did not want Venezuela to possess a great oil industry even after reversion to the nation takes place."[17] What seemed clear was the Pérez Alfonzo already had a deep distrust in the ability of Venezuelans to use the oil wealth sensibly and had started to believe that the best course of action was to reduce the relative importance of the oil industry in the national economy.

Defense of Oil Prices

The Betancourt government inherited a very unsatisfactory situation regarding the prices of Venezuelan oil in the world markets. The glut which had driven Middle Eastern oil prices downward in 1959 has already been mentioned. Although Venezuelan oil prices were higher and better protected than those of Middle Eastern producers due to the stability of the U.S. markets and the transport premium that Venezuelan oil enjoyed, they also suffered a decline in 1959 and later years (see table 2-1). Furthermore, the difference between these posted prices for Venezuelan oil and actual realized prices increased dramatically in 1959, and continued increasing in later years, going from 9 cents in 1956 to 88 cents in 1969. According to Danielsen the imposition of mandatory controls on imports ordered by President Eisenhower in 1959 resulted in a substantial reduction in the demand for oil from foreign countries, exerted greater downward pressure on world oil prices, and increased capacity utilization rates in the United States.[18]

In Venezuela the increasing concern over this situation led to the creation in 1959 of a group of energy and economic experts called the Coordinating Committee for the Conservation and Commerce of Hydrocarbons. This unit was part of the Ministry of Mines and Hydrocarbons. It monitored the sales of Venezuelan crude oil and products made by the foreign oil companies to international clients. The work of this committee proved to be useful and produced many upward revisions of the tax amounts paid by the companies in respect to those sales. In general, however, the most important attitude taken by the ministry was to oppose production increases, on the basis that larger volumes would inevitably lead to lower prices. This conservationist attitude of the Ministry of Mines and Hydrocarbons, however, was not shared by other government agencies, especially the Finance Ministry. As a result output generally grew and prices declined all throughout the 1960s.

Table 2-1
Posted Price Trends, 1950-1971
($/barrel)

Date	Venezuela, 35° API	Iran, 34° API	Saudi Arabia, 34° API
November 1950	2.570	——	1.750
February 1953	2.820	1.860	1.930
June 1957	3.050	1.990	2.080
February 1959	2.900	1.810	1.900
August 1960	2.800	1.810	1.800
February 1971	2.725	2.170	2.225

Source: OPEC, Annual Statistical Bulletin, 1976 (Vienna, 1977).

Preferential Hemispheric Treatment

In 1959 Dr. Pérez Alfonzo went to Washington to ask for special considerations for Venezuelan oil. He believed that Venezuela should be part of a hemispheric trade block, together with Canada, Mexico, and the United States in order to avoid unwelcome fluctuations in the price of oil. Pricing would be a matter of negotiation between governments rather than between the private oil companies and their clients. The Venezuelan government was especially worried about Mexico and Canada receiving preferential treatment over Venezuelan oil, since they felt that Venezuela had also special treatment earned for many years' service as a reliable supplier to the United States. Although Venezuela did obtain some minor concessions regarding increases in the quotas of residual fuel oils being sold to the U.S. East Coast, it never attained the desired preferential status.

Creation of a National Oil Company

In 1960 the government created the Corporación Venezolana del Petróleo (CVP), the Venezuelan Oil Corporation, a fully owned stated oil company. It was created as an appendix of the Ministry of Mines and Hydrocarbons to serve as an instrument of national oil policy. CVP was an integrated oil company with a board of directors, led by the minister of Mines and Hydrocarbons, which met at least once a month. The chief executive officer was the director general, who led an executive junta of four members. CVP was given prospective oil areas to develop as well as the plant and equipment required to commence activities.

The Venezuelan Oil Corporation was a latecomer to the arena of national oil companies. This was hard to understand in an oil-producing country that for many years had been the leading world oil exporter.

Ruben Sader Pérez, a former director general of CVP, defined the main objectives of the company as follows:

1. Participation in all phases of the industry
2. A greater participation in the benefits of oil exploitation

Sader Pérez went on to add,

> We constituted CVP with directors who had neither sufficient technical nor managerial preparation. . . . Unfortunately, to this date, we cannot say that there exists, either in the private or in the public enterprises, Venezuelans with sufficient training or in enough numbers to take care . . . of our petroleum development.[19]

Sader Pérez's comments illustrate the basic shortcomings of CVP. This company was formed to participate in all phases of industry, but it was never given the means to achieve that objective. CVP did become a rather efficient state-owned enterprise, at least during the period 1964–1969. It did some exploration work, developed a modest production that never went above 100,000 barrels per day, operated a small refinery and two oil tankers, and in general behaved like a small oil company, as compared to the much larger multinationals. In spite of the efforts of valuable people such as Sader Pérez and F. Delon, CVP never became what the political leaders who created it had envisioned: the national oil company. By the time nationalization came, it was clearly not the organization that could take over the industry in the name of the state. It was in fact a largely demoralized and mediocre corporation, already showing signs of a deep process of politicization and would eventually be merged with several other organizations as its only practical means of survival.

The Creation of OPEC

The concern of the Venezuelan government for declining oil prices and the pioneering work done during the years on the profit-sharing mechanism had won the respect of other oil-producing countries which shared the same concerns. Following the price reduction decided by the oil companies in 1959, producing countries met in Cairo on the occasion of the First Arab Petroleum Congress. Venezuela and Iran attended as observers. Pérez Alfonzo of Venezuela and Tariki of Saudi Arabia made a gentleman's agreement which would bring the main producing countries together in their efforts to prevent a further erosion of world oil prices. After a second round of price decreases by the major oil companies, in August 1960, the producing countries lost no time in responding. In September 1960 Iran, Saudi Arabia, Venezuela, Iraq, and Kuwait met in Baghdad and created OPEC.

> OPEC was immediately successful in freezing posted prices but not in restoring actual market prices to previous levels. . . . Actual market prices thereafter bore no relation to posted or tax-reference prices. Under these circumstances, the countries received a constant amount per barrel regardless of the market price . . . each country pressed its concessionaires to raise lifting levels even further. The result was greatly expanded crude oil production and continually lower market prices throughout the 1960s.[20]

Even after the departure of Pérez Alfonzo from the Ministry of Mines and Hydrocarbons, the oil policy of Venezuela continued unchanged. President Betancourt was replaced by President Raúl Leoni, also from Acción Democrática, for the period 1964–1969.

Leoni's administration advanced a series of negotiations with the oil companies on the issue of tax claims. The basis of these claims was the apparent decrease in national income derived from the difference between the actual sales prices and the prices declared by the companies for tax purposes. The estimates of the government of what prices should be were much higher than the sales prices declared by the companies. This difference was estimated by the government at some $600 million.[21] The final agreement, reached in 1966, provided in payments from the companies to the government of the order of $259 million.[22] What was important about these settlements, besides the large amounts of money involved, was that they illustrated the degree of sophistication that had been attained by the Venezuelan technical and financial experts monitoring the industry.

In 1966 the government also agreed with the oil companies for a five-year term on the prices to be utilized to calculate tax payments. This agreement was incorporated in the income tax law and survived until December 1970.

The U.S. oil imports policy continued to be a source of great tension between the United States and Venezuela. In 1969 President Nixon decided to create a cabinet task force to report on the oil import situation. This task force recommended a tariff system, instead of the quota system, a preferential treatment for Venezuelan residual fuel oils (30 ¢/barrel less duty than the barrel from other sources) and generally lower duties for oil imports from the western hemisphere. Unfortunately for Venezuela the recommendations of this task force led by George Shultz were harshly criticized by oil industry, coal, banking, and other local U.S. groups. In particular, critics felt that the oil-supplying capability of the western hemisphere had been overestimated. As a result of these criticisms the recommendations of the Shultz report were not accepted by President Nixon.

Meanwhile, CVP, the Venezuelan Oil Corporation, was having growth problems. Unable to become a production and refining giant like the multinationals working in the country, it had emphasized exploration and local marketing activities. At a time in which no exploration was being done by the multinationals, CVP conducted seismic surveys in Lake Maracaibo, in the Gulf of Venezuela and in vast land areas of the country. This effort, although important, was short of the levels really required by the country. The local market had been partly assigned to CVP through a decree issued in 1964 with the idea of giving this company eventual total control of the domestic market. Accordingly the participation of CVP in this market grew steadily from 3 percent in 1964 to 32 percent in 1971.[23] By the year of nationalization, 1975, this participation was over 50 percent. Although the participation in the domestic market gave CVP a welcome knowledge of a sector of industry, it was far too narrow and unprofitable an experience to help the corporation to become a really integrated oil company.

Of more interest to CVP were service contracts for association between the nation and foreign companies, which would replace concessions. The new type of association was announced by the government in 1961, but it was not before 1968 that explicit regulations about these types of agreements were issued. These regulations, called Minimum Bases for Service Contracts, were given to publicity by CVP in March 1968 and five blocks of 250,00 hectares each in South Lake Maracaibo were offered to oil companies for exploration and eventual development.

Service contracts became a very controversial issue. The political left attacked them vigorously as "concessions in disguise." They claimed that CVP would not have, through them, a real control of the industry's operations. Economist Maza Zavala expressed this view with much force when he said, "I consider that the best alternative for the development of the Venezuelan petroleum industry and the true instrument of the no more concession petroleum policy is the development, in all its aspects and phases, of an independent national oil industry represented by CVP."[24] For these political leaders there was no other alternative than total national control. On the other hand oil companies found the conditions imposed by the government too harsh and possibilities to make a profit almost nonexistent. Service contracts included, among others, the following stipulations:

A 3-year exploration period.

A 20-year exploitation period.

The contractor would keep 20 percent of the original area and CVP would retain 80 percent.

Ten percent of the oil production would go to CVP. The remaining 90 percent of the oil would be sold by the contractor, and tax and profit-sharing mechanisms would be applied which would give the nation a very large share of the profits.

All exploration investments would be of the sole responsibility of the contractor.

There would be operational and administrative control of the programs by CVP.

Oil Minister Pérez La Salvia estimated that the five South Lake Maracaibo blocks put up for bidding contained recoverable oil reserves of some 3.3 billion barrels. About eighteen companies submitted offers for the blocks. Shell obtained one block, Mobil obtained one block, and Occidental Petroleum obtained the remaining three blocks. In 2 years, from 1971 to 1973, about eighteen exploratory wells were drilled. None found what could be really called a commercial accumulation. The best-looking geological structure, located in the Shell block, proved to be totally dry.

Although the lack of success in the exploration of South Lake Maracaibo was very disappointing and robbed Venezuela of its hopes of substantially increasing oil reserves, service contracts served to evaluate an important portion of the Maracaibo Basin at no cost to the government. By this time the international events starting to take place in the Middle East and North Africa would introduce such a drastic change in the Venezuelan oil scene that service contracts would no longer be considered a politically valid option.

The 1960s ended in Venezuela with the inauguration of a new president. For the first time in Venezuelan political history, a candidate of the opposition, the Christian Democrat (COPEI) Rafael Caldera, was elected. He had to deal with U.S. oil-import regulations and made an early trip to Washington, in June 1970, to plead his cause, but without much success.

> What he left with was a 35,000 barrel a day increase in the U.S. quota on home heating oil and President Nixon's observation on the need for a hemispheric policy on oil. Until early 1973, the Caldera government believed that the United States would eventually recognize the justice of (its) position and of the advantages which hemispheric treatment would have for both countries. Instead it received more piecemeal concessions on various petroleum products.[25]

The increasing frustration of the Venezuelan political sector over this unsatisfactory situation contributed greatly to the wave of nationalism which swept the country in the early 1970s.

Notes

1. D.E. Blank, *Politics in Venezuela* (Boston, 1973), p. 19.

2. For a detailed description of the contents of this law, see L. Vallenilla, *Auge, Declinación y Porvenir del Petróleo Venezolano* (Caracas, 1973).

3. Reversion of assets meant all oil company assets would revert to the nation, without compensation, by the end of the concessions.

4. E. Owen, "Trek of the Oil Finders," American Association of Petroleum Geologists Special Publication (Tulsa, Okla., 1975).

5. Larson et al. (1971), quoted by Owen, ibid., p. 482.

6. In Owen, ibid., p. 1090.

7. Rómulo Betancourt, *Venezuela, Política y Petróleo* (Caracas, 1978), p. 289.

8. Ibid., p. 283.

9. Betancourt, *Venezuela, Política y Petróleo,* p. 282.

10. Betancourt, *Venezuela, Política y Petróleo,* p. 287.

11. Vallenilla, *Petróleo Venezolano,* p. 205.

12. Ibid., p. 206.

13. Quoted by E. Lieuwen, *Petroleum in Venezuela, a History* (London: Russell and Russell, 1967), p. 105.

14. Stephen Rabe, *The Road to OPEC: United States' Relations with Venezuela, 1919–1976* (Austin: University of Texas Press, 1982), p. 124.

15. Ibid., p. 128.

16. D. Bohi and M. Russell, *Limiting Oil Imports: An Economic History and Analysis* (Baltimore: John Hopkins University Press, 1978), p. 27.

17. Vallenilla, *Petróleo Venezolano,* p. 310. However, in the English version of this excellent work, *Oil, the Making of a New Economic Order* (New York: McGraw-Hill, 1975), Vallenilla modifies his evaluation of Pérez Alfonzo's motivations when he says, "In attempting to analyze the reasons that he might have had for seeking a decline in the oil industry and for efficaciously aiding that decline, we should take into account his point of view. For example, there is his animosity—in some cases, based on sound reasons—toward those who had retained, and still retain, control over the oil industry in his country. It is difficult, however, to conclude that his animosity in this respect constituted his basic reason for accentuating restrictions on the oil industry." (p. 118)

18. Albert L. Danielsen, *The Evolution of OPEC* (New York: Harcourt Brace Jovanovich, 1982), p. 150. Closed production in the United States dropped significantly from 1958 to 1959 and contributed to keeping imported oil cheap. However, it made the country much more vulnerable to a shortage in the world market, such as the one experienced during the 1973–74 oil "crisis."

19. Ruben Sader Pérez, *The Venezuelan State Oil Reports to the People* (Caracas, February 1969), p. 38.

20. Danielsen, *Evolution of OPEC,* p. 128.

21. Juan Pablo Pérez Alfonzo, *Oil and Dependency* (Caracas, 1971).

22. Memoirs, 1966, Ministry of Mines and Hydrocarbons.

23. Vallenilla, *Petróleo Venezolano,* p. 322.

24. Sader Pérez, *The Venezuelan State Oil,* p. 170.

25. Rabe, *The Road to OPEC.*

3

Accelerated Changes, 1970–1973

Libya Leads the Way

A turning point in the balance of the power relationship between oil companies and producing countries, it is generally assumed, was the Libyan revolution in 1969. Yet Libya under King Idris had already started to lead the way that would end, in 1974, with a significant reversal of the power roles being played by OPEC and the multinational oil companies. Libya had been the first OPEC member country to try to enforce the decisions of the eighteenth OPEC meeting held in Baghdad in November 1968, including the establishment of a very elaborate set of controls to be applied by member countries to the operations of the multinational oil companies. Through these controls the Libyan government would have acquired a determining influence in the programming of the production of each concessionaire. The concessionaires protested so vigorously, however, that the Libyan government agreed to make the regulations more palatable. It was while this revision was being made, in September 1969, that Colonel M. Qadhafi came to power by means of a bloodless military coup. There is little doubt that the Libyan initiatives regarding increasing controls over the operations led the Venezuelan government to issue some time later, in December 1971, decree 832, which took this producing country one important step further in the systematic administrative takeover of its oil industry.

The new Libyan government did not seem to favor a radical change in the relationship with the oil companies. The new prime minister said,

> There will be no spectacular changes in our oil policy and I can confirm that we shall endeavor to cooperate with the oil companies, provided that the interests of the Libyan people . . . are taken into account. It is possible to safeguard these interests by means of a more effective control over oil operations.[1]

However, a few weeks later, Qadhafi told oil company representatives,

> The Libyan people, who have lived for five thousand years without petroleum, are able to live again without it.[2]

35

Meetings between the Libyan officials and the oil companies centered around the desire of the Libyan government to increase posted prices. In April 1970, however, Amoco posted prices for a field newly opened to production which did not satisfy the desires of the government. The Libyan government retaliated by utilizing the OPEC-agreed regulations of November 1968 to order a cut in production. The reason given for the order was that "the companies were producing oil wells at above their most efficient rate."[3] Occidental Petroleum received the order and, in late May, their output went down from 800,000 barrels per day to 400,000. Since this company was unable to obtain additional oil at cost from other oil companies, it agreed, in September 1970, to increase prices of Libyan oil by 30 cents per barrel.

This increase is generally taken by analysts of the Libyan oil industry as marking "the end of an era in the international oil industry and the beginning of a new one. The creation of OPEC in 1960 had blocked any downward movement in posted prices throughout the world after that date. The oil companies themselves had blocked any upward move."[4] Although the initiative was largely the product of Libyan political intuition, it was effectively aided by the closing of the Trans-Arabian pipeline in May 1970 and the resulting increase in tanker rates, which put a premium on Libyan oil in the European and U.S. markets.

Once Occidental agreed to a price increase, the other multinational oil companies operating in Libya followed suit. Other oil-producing countries in the area made similar demands, and the power to increase oil prices was transferred into the hands of OPEC. The Venezuelan National Congress had been following the Libyan events with particular attention, and the opposition parties pressed the government to take advantage of the situation. Oil Minister H. Pérez La Salvia appeared before the Senate and admitted that in "situations such as this, of a general price increase due to rising demand . . . Venezuela . . . is not profiting from the increasing prices paid by the consumers."[5]

This realization led the political parties of the opposition, especially Acción Democrática and Union Republicana Democrática, to propose and push through a partial reform of the income tax law establishing a tax tariff of 60 percent for oil companies and authorizing the National Executive to fix the export reference prices unilaterally, to replace the existing system of agreed reference prices, which had been established during the 1960s. This was a decision of the greatest importance, since it would help to make possible for the government to predict, almost to the last cent, what their oil income would be in any given year. This was accomplished with decree 832, which stipulated that all exploration, production, refining, and sales programs of the oil companies had to be approved in advance by the Ministry of Mines and Hydrocarbons. The government now knew in advance what the level of production would be and could calculate the amount of income tax the foreign oil companies would have to pay.

OPEC Meets in Caracas

In December 1970 OPEC held its twenty-first meeting in Caracas under an intense climate of nationalism which made possible the passing of a series of resolutions. OPEC resolved

> To establish uniform criteria among member countries to calculate the value of the different crude oils being produced in OPEC countries.
>
> To improve fiscal participation of the member countries on the basis of a minimum tax level and to establish reference prices unilaterally, just what the Venezuelan government had decided.
>
> To agree on production cuts if such a decision was required to stabilize oil prices.
>
> To act in unison and to exercise fully the rights of the member countries whenever this was required.

After this meeting "Every OPEC member, with the exception of Indonesia, either made public statements or (more convincingly) told the companies privately that if their demands were not met, all oil production would be stopped and the companies would then have to face the wrath of the consuming countries."[6] OPEC spokesmen adopted a new, challenging tone, very much aware that they had found weak spots in the position of the oil companies and the consuming countries.

The Venezuelan Government Increases Control

In March 1971 President Caldera announced that Congress would be asked to pass a law to reserve to the state the utilization of natural gas. The draft of this law had been prepared by an important group of lawyers with a sound experience in hydrocarbon issues. When the project reached National Congress, however, it became subject to strong criticism from the opposition, who considered its terms too moderate. In particular they rejected the idea of a compensation to concessionaires for the gas to be delivered to the state. The gas, they claimed, should be delivered at cost.

The opposition also rejected the provisions included in the draft bill for the eventual creation of joint gas-exploitation ventures between state and private companies. Another important criticism of the draft bill had to do with the existing liquified gas projects. The opposition refused to allow the utilization of nonassociated gas for this purpose. A law nationalizing the natural gas industry was finally passed in August 1971. It was not particularly significant in practical terms, although it helped to reinforce the deeply nationalistic climate existing in the country.

Of much more importance was the passing of the law of reversion. This was a bitterly fought issue. The debates over this law were the forerunners of the great debate about nationalization to ensue a few years later. On the basis of the reversion law, passed in July 1971, all the assets, plant, and equipment belonging to the concessionaires within or outside the concession areas would revert to the nation without compensation upon expiration of the concession. The law also required the concessionaires to deposit funds in Venezuelan banks to guarantee that the assets would be kept in good working condition for the life of the concession.

Several of the oil companies operating in Venezuela—Creole (Exxon), Shell, Texas, Mobil, Mene Grande (Gulf), and Sinclair—went to the Venezuelan Supreme Court asking for the nullification of the law, on the grounds that some of its clauses violated their constitutional rights and were of a confiscatory nature. In turn the Venezuelan Attorney General J.G. Andueza introduced before the Court in March 1972 a document in defense of the law. Andueza particularly defended the rights of the nation to include in the reversion law those assets of the concessionaires located outside concession areas.

Feelings ran high on both sides. In general, oil company executives and business groups such as Fedecamaras argued that the law seemed to change the rules of the game, since the assets located outside the concession areas such as headquarters buildings would probably never have been built if it had been known that they would be taken over by the State. The argument of the state was that all assets built or acquired by the companies, directly related to the operation of the concessions, should be subject to reversion, regardless of where they were located.

The appearance of the law of reversion and decree 832 coincided with an almost immediate and significant reduction in the Venezuelan oil output. Production decreased by about 5 percent in 1971 and by more than 10 percent during the first quarter of 1972. According to an analysis prepared by the Venezuelan Central Bank this reduction could be explained by lower tanker rates, which gave Middle Eastern oil the edge over Venezuelan oil. The analysts also mentioned a mild winter as a contributing factor in lower demand and hence the lower output. An analysis issued by the Ministry of Mines and Hydrocarbons made some contradictory claims, however: that there were ''no valid reasons for the production cuts of 1971 and, much less, those of the first months of 1972. . . . Venezuelan oil could have been sold in the main markets during 1971 in equal or better terms than those coming from other areas."[7]

The political sector took up the theme and several congressmen accused the oil companies of retaliation. The president of the Committee of Mines and Hydrocarbons of the lower chamber, F. Faraco, denounced the companies for "closing down 700 wells and reducing the output in 635,000 barrels per day," adding, "We are witnessing a maneuver against the country."[8] President Caldera, speaking to a World Meeting of Oil Workers held in Caracas in February 1972, suggested that the production cuts were artificially provoked. He said,

> World oil demand increases every day, but Venezuelan oil output decreased last year. . . . Although there are natural causes which influenced this decrease . . . some rumors try to present it as an indication that the nationalistic position of Venezuela is being opposed by a position which could progressively close markets to our oil.[9]

Oil Minister Pérez La Salvia emphatically said,

> that they increase (demand) elsewhere cannot be explained with purely economic reasons . . . there must be noncommercial reasons, among which must be the displeasure of those companies with the measures Venezuela has taken in connection with our main source of wealth.[10]

The *Washington Post* was reported by the Caracas press to comment on the Venezuelan oil situation as follows:

> Venezuela seems to be preparing for a nationalization of . . . the petroleum investments . . . The U.S. government is very worried not only about the growing volume of U.S. investments (in Venezuela) but also about the growing U.S. dependence on Venezuelan energy supplies.[11]

The Evening Star, an afternoon Washington daily, was also reported as saying:

> a long period of friendly business relations seems to be coming to an end . . . due to the differences between the oil companies and the Venezuelan government . . . if they do not start talking the same language soon, the two countries will eventually collide.[12]

The 1972 annual report on U.S. foreign policy issued by Secretary of State William P. Rogers said of U.S.–Venezuelan relations:

> Oil problems kept in the forefront of U.S.-Venezuelan relations during 1971. . . . Consultation was less fruitful than during the preceding year. . . .

> The increasing volume of our requirements will put to the test the supply
> capacity of the Western hemisphere producers, especially Venezuela, where
> any significant (production) increases will require substantial amounts of
> new investments. A growing proportion of our imports, therefore, . . . will
> have to come from more prolific sources.[13]

This statement created quite a stir in Venezuela.

The Venezuelan Embassy in Washington, D.C., made public a document claiming that the country had a shut-in capacity of about 700,000 barrels per day and considerable heavy oil reserves in the Orinoco area.[14] This document was in answer to Rogers's document and, also, to the opinions given March 21 by Adjunct Secretary of the Interior Hollis Dole to a Committee of the U.S. House of Representatives. According to Dole, Venezuela could not be expected to supply the growing import requirements of the United States since its supply capacity had already peaked.

Venezuelan political leaders gave much weight to Rogers's statements. For weeks the Venezuelan press was filled with voices of concern. Acción Democrática's president, Gonzalo Barrios, warned against overreaction when he said, "There are no motives for a general mobilization of public opinion in defense of our national petroleum policy. An exaggerated reaction . . . could lower the credibility on our seriousness of purposes."[15]

The Venezuelan public also gave attention to the law of reversion and other nationalistic measures taken by the government. The oil companies, represented by executives such as Guillermo Rodriguez Eraso, Ernesto Sugar, Kenneth Wetherell, Wolf Petzall, R.N. Dolph, Luis Guillermo Arcay, and Pedro Mantellini went to Congress to express their views on the law. They claimed that although the concept of reversion was impeccable, the law went far beyond the original scope of the concept. In those claims the oil companies were accompanied by most Venezuelan business groups, including the National Economic Council, Fedecámaras, the Venezuelan oil company "Mito Juan," the Caracas Chamber of Commerce, and others. In favor of the law were practically all of the political parties in the country. Once the law was passed the political sector saw the reduction in the Venezuelan oil output as a retaliatory move on the part of the companies.

With the debut of the law of reversion and decree 832, the relationship between the Ministry of Mines and Hydrocarbons and the oil companies changed significantly. For many years the ministry had essentially been a fiscalizing agency in charge of ascertaining that the requirements of the law of hydrocarbons were being met. Since 1959 the ministry had added a monitoring function on the prices of crude oil and products being sold by the companies in the world markets to make sure that the government was obtaining a fair share of the earnings. The activity had always been one of verification after the fact. But now its nature changed significantly. The convergence of all the new tools: the law of reversion, decree 832, the law of natural gas, the authority to fix unilaterally the fiscal export prices, gave government almost total control of the oil industry practically at no risk. For all practical purposes the Venezuelan oil industry was in the hands of the state by 1972. The ministry staff which, up to then, had been mostly auditors of the industry now became comanagers. There was nothing that the industry could do without the previous approval of the ministry. The companies had to submit to the ministry, on the basis of decree 832, a detailed program of activities including exploration, production, refining, local sales, international sales, and all financial data. The ministry staff would receive these documents and could suggest minor or important changes in the programs. The number and locations of exploration wells could change, production levels or refining levels could be modified, clients for the oil could be vetoed, prices could be challenged. In addition, through a combination of its new managerial powers and its authority to fix fiscal export prices, the government could now make sure that it had all the income it required for the national budget.

Once the strategic decisions were no longer fully in the hands of company management and the constraints imposed by government on the concessionaires became so great, motivation within the oil industry's organizations decreased considerably. At the same time some ministry staff became rather arrogant in dealings with company representatives. The fact that many of these company representatives were Venezuelans did nothing to decrease the tension. In fact, it aggravated it. The letters from ministry bureaucrats to the companies were frequently written in a cold, even rude style irritating to their Venezuelan addressees. The relationship between the ministry staff and the Venezuelan technocrats working for the oil companies rapidly became one of rivalry and mutual dislike, setting the stage for the confrontations which would come later, both during and after the process of nationalization.

The impact of this deteriorating relationship between the ministry staff and the oil-company managers cannot be sufficiently stressed. After nationalization became the most powerful single ingredient in shaping the relationship between the ministry and Petróleos de Venezuela (PDVSA), the national oil holding company. With nationalization, in 1975, there was no further apparent reason to keep, within the ministry, the comanagement of the industry that had been typical of the transition period 1972–1975, from the moment the law of reversion was passed until nationalization. But although conditions were clearly different after nationalization, the attitude of ministry staff was not. They remained adept at exercising managerial power. They insisted, for example, that decree 832 was still valid and kept asking the nationalized industry for their operational programs to approve or disapprove them in advance. Although Petróleos de Venezuela had been created in August 1975 precisely for that purpose, ministry staff insisted on dealing with the nationalized oil industry as if nothing had changed. For several years after nationalization, concerned analysts of the oil situation kept asking for a clear definition of the ministry and PDVSA's roles as a cure for the ever-present tension between the two organisms. But the friction had never been a product of lack of clarity on each side about their respective roles; rather, it was the product of the reluctance of the ministry staff to accept their true role. Although they knew well that their role was one of fiscalization and policy-making, they insisted on trying to manage the industry. When the industry objected, open confrontations took place.

The attitude of the ministry staff was not hard to understand. For many years they had been overworked and underpaid. They were monitoring an industry which paid its managers and engineers considerably larger salaries than they, the bureaucrats, earned and gave them housing, transportation, and other benefits that government bureaucrats could not enjoy. It was inevitable that some of the staff would end up projecting their dissatisfaction with this state of affairs toward the people working for the companies. They saw these people as mercenaries working for foreign owners, whereas they felt they themselves worked for their country. In time they came to see themselves as true nationalists and the company managers and engineers as people who were not to be trusted. When nationalization took place and the same managers and engineers who had been running the foreign companies became the key staff of the new nationalized organizations, the ministry staff felt that nothing had changed, that the foreign companies were still in command through their former subordinates. The attitude of some political parties tended to reinforce this feeling of distrust and to encourage the notion that the ministry staff and the personnel of CVP, the original Venezuelan state oil corporation, were the only truly patriotic oil-industry employees.

The increasing power of the ministry staff had some other consequences. The ministry was not, it has never been, a very cohesive organization. The different offices, or divisions: hydrocarbons, reversion, local market, petroleum economics, geology, were very compartmentalized and had little interaction. During the years 1972–1975, each division tried to increase its power, frequently at the expense of other departments within the organization. The new Reversion Division, created in January 1973, started a very active drive for information from the oil companies, most of which already existed in the files of the Hydrocarbons Division. The rivalries between the two division heads, Calderon Berti and Reyes, made impossible a true cooperation between these groups. As a result the oil companies had to spend thousands of hours duplicating information and the general climate of the relationship deteriorated even further. By 1973 the bureaucratic pressures on the oil company staff and managers were so great that nationalization started to sound to them like a reasonable idea.

The 1973 Crisis

In 1972 the oil-producing countries of the Persian Gulf flexed their new-found muscles by establishing negotiations with the oil companies leading to ownership participation. Although they rapidly obtained a 25 percent participation, less than a year later they revised those agreements to obtain up to 60 percent participation in the ownership of the companies. In 1972 and 1973 oil prices went up by about 35 cents per barrel. In early 1973 crude-oil and product inventories in the United States were very low and there was open talk of an impending energy crisis. President Nixon ordered a study on "the relationship between energy policies and foreign and security concerns," to a task force consisting of H. Kissinger, G. Schultz, and J. Ehrlichman.[16] In April 1973 Nixon replaced the import quota system by a license system, to gain flexibility to increase oil imports. Kissinger favored the development of bilateral ties with Saudi Arabia and claimed that major progress had already been achieved in that direction.[17] In October 16 OPEC Persian Gulf states members decided to raise their posted prices by 70 percent and to cut production progressively until Israel retreated from the territories it had occupied during the war which had started October 6. In addition, an embargo was imposed on countries friendly to Israel (the United States and Holland). OPEC met in Tehran December 1973 and raised prices again, this time from an average of $5.12 per barrel to $11.65 per barrel. "Within 48 hours the oil bill for the United States, Canada, Western Europe and Japan increased by $40 billion a year; it was a colossal blow to their balance of payments, economic growth, employment, price stability and social cohesion. . . . All the countries involved, even the producers themselves, faced seismic changes in their domestic structures."[18]

Although Kissinger had favored a bilateral arrangement with Saudi Arabia, the U.S. State Department in July 1973 denied Venezuela the possibility of a government-to-government arrangement for the development of the heavy-oil reservoirs of the Orinoco area on the grounds that "bilateral deals would set off a bidding war, pitting one consuming nation against another."[19] This inconsistency in the U.S. oil policy seemed to stimulate Venezuelan nationalism even more. In 1973 the presidential candidate of COPEI, Dr. Lorenzo Fernandez, announced his proposed government program including the concept of an early reversion, whereas Oil Minister Hugo Pérez La Salvia, in a public meeting held in Maracaibo pointedly remarked that "the oil-producing countries have reached the conclusion that their (oil) industries should be in their own hands."[20]

Notes

1. F.C. Waddams, *The Libyan Oil Industry* (Baltimore: John Hopkins University Press, 1980), p. 229.

2. Ibid., p. 230.

3. Ibid., p. 232.

4. Ibid., p. 233.

5. Luis Vallenilla, *Auge, Declinación y Porvenir del Petróleo Venezolano* (Caracas, 1973), p. 473.

6. J.E. Akins, "The Oil Crisis: This Time the Wolf Is Here," *Foreign Affairs* (April 1973):472.

7. *El Universal,* March 22, 1972.

8. *El Universal,* March 7, 1972.

9. *El Universal,* February 26, 1972.

10. *Ultimas Noticias,* February 17, 1972.

11. Reported in *El Nacional,* February 26, 1972.

12. Reported in *El Nacional,* March 7, 1972.

13. Reported in *El Nacional,* March 8, 1972.

14. *El Universal,* March 24, 1972.

15. *El Nacional,* March 11, 1972.

16. H. Kissinger, *Years of Upheaval* (Boston: Little, Brown, 1982), p. 869.

17. Ibid., p. 871.

18. H. Kissinger, *Years of Upheaval* (Boston: Little, Brown, 1982), p. 885.

19. Quoted by S. Rabe, *The Road to OPEC: United States' Relations with Venezuela, 1919–1976* (Austin: University of Texas Press, 1982), p. 177.

20. Quoted by Irene Rodriguéz Gallad, and F. Yanez, *Cronología Ideológica de la Nacionalización Petrolera en Venezuela* (Caracas, 1977), p. 139.

4

The Public Oil Debate, 1974–75

Although the political campaign leading to the presidential elections of December 1973 was long and hard fought, oil issues were dealt with in a very prudent manner by both leading presidential candidates, Carlos Andrés Pérez of Acción Democrática and Lorenzo Fernández of the Christian Democrats (COPEI). Pérez promised a national debate before a decision on nationalization was made. Fernández was more aggressive and repeated frequently that his hands would not shake when the moment came to sign the nationalization bill. The mood of the country was definitely for restraint in this area as Venezuelans knew that the oil industry was the main, if not the only, means of national support.

The moderation of the silent majorities was not shared by the political left. The presidential candidate of the small socialist party, MEP, J.A. Paz Galarraga accused both Acción Democrática and COPEI of a secret agreement with the foreign oil companies to establish mixed companies and, even, to issue new oil concessions. Pompeyo Márquez, secretary general of MAS, a medium-sized socialist party, claimed that the act of nationalization was no longer revolutionary. Márquez proposed, in turn, the "socialization" of the oil industry, without describing in much detail what he meant. Although President Caldera had stated in February 1973 that Venezuela had no intention of nationalizing the oil industry, Fernández, campaigning in March, enunciated a platform which included the "advancing of the act of reversion," a euphemism for nationalization. The lack of coherence in the government's position was confirmed by the Oil Minister H. Pérez La Salvia, who, speaking in the Seventh Journalists' Banquet, a yearly headline-making event, said that the administration of the oil industry "could not be left in the hands of the international companies." One week later Pérez La Salvia added that the word "nationalization" should not scare anyone since this was a concept accepted in the Venezuelan constitution and laws. The finance minister, L.E. Oberto, seemed to argue against Pérez La Salvia when he said that the decision to nationalize the oil industry should not respond to capricious reasons. A few days later Pérez La Salvia reiterated that the government "was planning to nationalize the industry since its personnel was already fully Venezuelan and markets for Venezuelan oil were abundant."[1]

It seems highly possible that the insistence of Pérez La Salvia in stating that the government would nationalize the oil industry contributed significantly to the defeat of COPEI in the elections of December. Even some highly placed ministry oil experts like Ramsey Michelena, director of the Petroleum Economics Bureau, went on record to say that nationalization would be very risky "if a careful study of all alternatives was not previously made." In a paper presented to the Venezuelan Congress of Economists, he stressed that even new concessions could be a valid alternative as the government already had almost total control over operations, finance, and exports.

Acción Democrática's candidate Pérez adhered quite consistently to a line of moderation. He appeared on television in August 1973 to say that if he won, he would reach a decision based on a detailed study of the issue, incorporating the "fundamental sectors of public opinion."

The presidential victory of Carlos Andrés Pérez was surprising in its magnitude. He won by a landslide over his adversary Lorenzo Fernández, who accused the foreign oil companies of plotting his defeat. One day before the election, the Communist newspaper *Tribuna Popular* had published the photostat of a check for Bs 500,000 allegedly signed by the vice president of Shell, Alberto Quirós, to the electoral bank account of C.A. Pérez. This check was proven to be a forgery and did more harm than good to the cause of COPEI.

During the lame-duck end of term of Caldera, the pressures to initiate the execution of nationalization increased considerably. The government received unexpected support from Enrique Tejera París, a member of Acción Democrática and former Venezuelan ambassador to the United States, who declared, on December 28, that "the nationalization of the foreign oil companies should be a matter of immediate consideration." Caldera took up this theme rather enthusiastically and Pedro Pablo Aguilar, secretary general of COPEI, inquired if the Tejera París statement represented the official position of Acción Democrática. The defeated candidate, Lorenzo Fernández, sent a telegram to President Caldera asking him for the immediate nationalization of the oil industry. Pérez La Salvia joined in the attack stating that "all studies required to take this decision had been made by the Reversion Department of the Ministry." Acción Democrática rapidly disavowed Tejera París through the main party spokesman on oil matters, Arturo Hernández Grisanti. Meanwhile the presidents of both Shell and Creole, Kenneth Wetherell and Robert Dolph, suggested that their companies were ready to negotiate with the Venezuelan government on the basis of any reasonable new scheme of relations. The companies had already seen that nationalization was just a matter of time. Their financial returns were extremely poor in the second half of 1973 due to the increasing difference between tax reference prices and actual prices. Shell had shown a loss of about two U.S. cents per barrel during this period. Poor financial returns

and the increasingly constrained atmosphere under which the oil industry
was working were probably determinants in the process which ultimately led
to a negotiated nationalization. In early 1974 the government, through Minister Pérez La Salvia, continued to maintain that all studies necessary to execute the nationalization were ready. In the first week of January, Creole
was ordered to return to the nation the Jusepin and Mulata oil fields of
Eastern Venezuela, on the grounds that these fields had not been properly
exploited during the last years. Creole tried to take this case to the Supreme
Court, without success. Pérez Alfonzo, from his retirement headquarters in
Caracas, backed the decision of the government and asked for an immediate reversion of all unexploited oil fields and even of those still producing,
"recognizing fair payment only for really productive areas." MAS accused
Acción Democrática and the government, once again, of having a secret
understanding with the U.S. State Department and the Venezuelan private
sector, represented by Fedecámaras, to establish the basis of a new petroleum relationship essentially involving the creation of mixed enterprises.

COPEI and MEP Make a Move

Every actor concerned was now certain that there would be no waiting until
1983 for the Venezuelan government to take over the oil industry. Moreover
everyone concerned, including the foreign oil companies, seemed willing to
take this course. All that remained was how to proceed. A significant hint
was given publicly by COPEI when its president, Godofredo Gonzalez,
revealed that Venezuela would be prepared to pay the foreign concessionaires the amount of Bs 5,559 million (about U.S. $1.3 billion) as indemnity
for their assets and that COPEI was getting ready to introduce in National
Congress a draft of a nationalization bill. A week later the Movimiento
Electoral del Pueblo, MEP, also announced that it would introduce a draft
of a nationalization bill "to reserve for the state all phases of industry."
Both documents went to Congress in March.

The MEP Draft Bill

The draft produced by MEP was by far the better one. It was largely the
work of Alvaro Silva Calderón, a brilliant economist who had also been the
driving force behind the reversion law. MEP's draft bill made clear that the
oil industry should become a fully state-owned enterprise and that nationalization "should not mean the turning over of oil assets from foreign private
hands into Venezuelan private hands." MEP added, "It is not a substitu-
tion of the foreign private activity" but rather, it will be designed

> to let the people take direct part in the handling of their hydrocarbon resources and obtain the . . . benefits . . . , a purpose which is not guaranteed by the exploitation of the industry by national private capitals since it is difficult—if not impossible—to determine when capital is really national or when it is from outside.

MEP went on to say that

> to leave in private hands an activity on which not only national economic life depends but also its security and existence is not in accordance with the missions of the state.

The underlying reasons given by MEP for the country to undertake the nationalization of the oil industry can be summarized as follows: The energy crisis had dramatically altered the existing power relationships between producers and consumers, making it easier for producers to take direct control of its resources. The immense majority of oil-industry personnel was Venezuelan, and foreigners could, if they wished, find a place in a national oil industry without fear of being an object of distrust for their former affiliation to multinationals. Financing had always been internally generated and there was no reason why it should not continue to be so. International markets existed and were more and more amenable to a government-to-government relationship. Economic profit margins in the oil industry were so large that only the inexcusable negligence of governments or the acceptance of total government incompetence could further delay the nationalization of this industry. The yearly profits of the concessionaires were estimated at no less than Bs 24 billion (U.S. $5.8 billion), an amount which would go directly to Venezuela's treasury.[2] The national economy needed to be integrated; up to then the oil industry had remained disincorporated to the rest of the economy and the country had been too dependent on one single sector. And finally, there was the compelling reason of sovereignty—of national security and dignity—to make Venezuelans the direct administrators of the oil business. Venezuelans should not listen to pretexts of immaturity or incompetence distracting them from the main objective.

The law was to be comprehensive. It would reserve to the state all oil activities, from exploration to marketing. It provided for compensation of oil fields in production but not for those abandoned or in uneconomic exploitation. Activities would be undertaken by totally state-owned companies. Supervision of the activities would be done by a hydrocarbons supreme council, made up of the ministers of Mines and Hydrocarbons, the minister of Finance, two members named by the president of the republic, and three members chosen by the National Congress. This organism would be accountable to the president and would be controlled by the National Executive (four of seven votes), although Congress would have a very strong voice. It would have the authority to designate officers and to create the most appropriate organization to administer the national oil industry. Congress would exercise control after the fact by receiving information from the council on oil-industry performance. Compensation could not be, in any case, greater than net book value. State oil companies would have to pay amounts to the national treasury equivalent to those the concessionaires were obliged to pay according to existing laws. A special fund would be created with 20 percent of oil income to invest in carefully selected development projects. Employees would enjoy total stability except at the executive level.

Although the contents of this draft bill now look moderate, at the time they created furor, not only within the oil industry but also in public opinion at large. The Venezuelan private sector bitterly attacked it because it gave them no role to play. Many oil-industry executives resented the aggressive attitude of MEP representatives, who took to the press to label dissenters to their draft as unpatriotic.

The COPEI Draft Bill

COPEI's draft bill had many points in common with the one presented by MEP. The document reserved to the state all oil activities and prohibited the creation of mixed companies. Concessionaires had to give back to the nation the concessions and all assets in those concessions in good working condition. Good working condition would be guaranteed by a special fund equivalent to the 10 percent of cumulative gross investments of the concessionaires, to be deposited in a state bank and to be progressively freed as assets were returned. Compensation would be calculated on the basis of net book value. The control of the nationalized oil companies would be in the hands of a federation of state oil companies directed by a council composed of the minister of Mines and Hydrocarbons, who would be the chairman, the ministers of Finance and Development, two representatives of the National Executive named by the president of the republic, three representatives of Congress, and one representative of the oil workers' union. This council would name the officers of the operating companies. Of special interest was an article including the proviso that manpower in the oil industry should not exceed, by more than 10 percent, the average levels of 1973.

Certain conceptual elements advanced in both draft bills became readily accepted by the political sector. These were as follows:

The law was to reserve to the state all oil activities.

Compensation to concessionaires would be based on net book value.

General overseeing of the operating companies would be the task of a supervisory council.

The stability of oil-industry employees would be protected.

Administrative structures of the nationalized oil companies would remain intact.

The political parties paid less attention, though, to some other equally important issues such as organization, financial mechanisms, managerial systems, technological capability, and marketing. To them, to pass a law seemed to be all that was needed. In fact a whole plan had to be articulated to regulate management of the nationalized oil industry. It was unlikely that the political sector would have the time, expertise, or even the inclination to formulate such a plan. At the same time there were few signs that politicians were thinking of asking the oil technocracy for their cooperation or support. On the contrary, in fact, there were indications that the political sector did not consider oil industry managers very trustworthy. After all, they thought, these people had been working for such a long time for foreign companies that their patriotism was perhaps uncertain. Oil nationalization was too serious a matter, politicians claimed, to leave it to the oil experts. The political parties were engaged in a nationalistic race, each trying to be more radical than the next. The Communist party, for example, was not in agreement with the payment of compensation. The small leftist party MIR (Movimiento de Izquierda Revolucionaria) backed the Communist position on the basis that "foreign oil companies had—for years—evaded income tax payments." The most important parties such as Acción Democrática, COPEI, and MAS agreed that compensation should be paid, however. Pro-Venezuela, a group of owners of small and medium-size businesses, intellectuals, and politicians which traditionally opposed Fedecámaras, went on record to say that the oil companies had done so much ecological damage to the country that they should pay for these damages instead of receiving compensation.

The Oil-Industry Employees Get Together

At this moment a small group of oil-company employees decided to play a role in the debate. They were alarmed to see that the nationalization bills had been drafted by politicians without the cooperation of oil managers or technicians. They were worried about the conceptual and even arithmetic mistakes in the calculations and arguments incorporated in the documents and thought that the outcome of the debate would be uncertain unless they participated. The first meeting was held in Shell's coffee shop and was attended by a handful of employees. They decided to prepare an analysis of the situation. This analysis included an estimate of probabilities of a nationalization taking place and an attempt at determining the interests of the different sectors involved, as follows:

Interests of the Nation
1. To obtain maximum income per barrel
2. To ensure optimum reservoir performance
3. To exercise full control over the industry
4. To own the assets of the oil industry and to be able to maintain them properly
5. To simplify operations
6. To maintain optimum operational, administrative, and technological standards
7. To ensure the required levels of investment

Interests of the Major Political Parties
1. To combine a nationalistic stance with the interests of the various groups within the establishment

Interests of the Federation of Oil Workers
1. To maintain the status and economic benefits of oil workers
2. To obtain total work stability
3. To attract substantial deposits for the Worker's Bank, an organism newly created by the union

Interests of the Foreign Oil Companies
1. To maintain a presence in Venezuela
2. To work under clear rules of the game
3. To obtain adequate compensation
4. To have access to new oil sources
5. Not to create precedents in their Venezuelan dealings which could negatively influence their position in other countries

Interests of Service and Engineering Companies
1. To obtain as much work as possible
2. To get paid in time
3. To be able to compete for the execution of oil industry projects

Interests of the Oil-Industry Employees
1. Career stability
2. To maintain traditional personnel administration procedures in the oil industry
3. To ensure that their legal severance payments would be respected

4. To be eligible for international cross-postings as a means to become better technicians and managers
5. To be able to reach through merit the highest executive positions in the industry

Interests of Local Capital
1. To obtain a greater participation in the oil business and its related activities[3]

The oil-industry employees believed that the final solution should satisfy most of these interests if it was to be effective. In trying to enter the debate the employees knew that there would be a high degree of inhibition on the part of many of their colleagues for fear of retaliation by their employers or by the government. They also knew that the political world would tend to look on them as foreign agents and that the media might be reluctant to give them equal space. Not many oil-industry employees felt at ease writing to a newspaper or appearing before television cameras. Most were shy; some were indifferent and tended to concentrate on their own small specialized worlds. Nevertheless the leading organizers believed that the employees should initiate the formation of a group which could influence in some way the nature of the final decision about the Venezuelan petroleum industry. They believed that such a group should have as its basic mission to cooperate with other groups to help the nation decide the best course of action in the issue of oil nationalization. They decided to call a meeting and invite all interested employees for a preliminary exchange of views. This meeting was held at the Tamanaco Hotel March 27, 1974 and attracted more than 250 employees, who contributed about Bs 5,000 ($1,200) to pay for the rented room and soft drinks. In this meeting employees argued that all national sectors were talking about oil issues except them and decided that a group composed of oil experts was needed to inform public opinion about the oil industry, about how it was run, and what were the problems involved in its possible takeover by government. The outcome of the meeting was to structure an executive committee made up of eight employees and agree on the basic objectives of the organization: to participate actively in the debate on the oil industry nationalization; to cooperate with public and private sectors in the analysis and evaluation of the alternatives for nationalization; to help the nation to reach a decision in an objective way.[4]

The initial meeting of the oil-industry employees and the creation of the Agrupación de Orientación Petrolera (AGROPET) caused a strong reaction among the Venezuelan political world, especially the political left. The next day Alvaro Silva Calderón, the oil spokesman for MEP, went to the newspapers to accuse the foreign oil companies of pressuring against nationalization through its employees and claimed that they had been transported to the meeting "in company airplanes."

The relatively substantial attendance to the Tamanaco meeting encouraged the employees to continue their efforts. Minister Valentín Hernández received them. A petroleum engineer by training, Hernández had not been an oil-industry employee. Although he had worked in the oil industry at the outset of his career, he soon went on his own and had made a small fortune drilling water wells. He was Venezuelan ambassador to Austria when Carlos Andrés Pérez won the presidency and called on him to serve as oil minister. It is said that Hernández's selection was suggested to Pérez by Rómulo Betancourt. An unknown quantity in Pérez's cabinet, he soon won the respect of the oil industry by his skillful handling of the difficult situations which often came up during the nationalization process.

The Nationalization Commission

In May 1974 President Pérez created the Nationalization Commission, a task force in charge of analyzing the mechanisms required to allow the National Executive to take over the assets of the oil industry before the normal reversion year of 1983. As he said in his speech:

> I have decided, with the unanimous backing of the cabinet and abiding by the mandate received from the Venezuelan people last December 9 (election day), to proceed immediately to rescue the oil concessions that would revert to the nation in the decades of 1980 and 1990, without waiting for the normal expiration date established for those concessions.

> the nationalization of the oil industry is not a rhetorical chapter, but a plan for action. The softer we speak, the easier it will be to listen and to understand one another. If we shout, we run the risk of . . . confusing our goals.[5]

Pérez went on to state that the previous oil minister, Pérez La Salvia, had not been right when he claimed that the Ministry had made all studies required to execute the nationalization. The president said: "There are not, in the ministry, any studies that include the multiple technical, economic and legal aspects which are needed to . . . nationalize. . . . The studies had not been made." In his speech, Pérez gave detailed terms of reference to the Commission, including determining a formula to calculate the amount of compensation to concessionaires, procedures to operate the industry during the transitional period, and procedures to guarantee the legal payments and funds of the workers.

He specified that compensation should not exceed net book value and should be paid in public debt bonds. The object of nationalization would be the assets and not the companies themselves. This meant that new corporations would have to be created. Pérez left little room for discretion on the part of the commission when he said: "There will be four companies created, to replace Creole (Exxon), Shell, Mene Grande (Gulf), and a fourth one to replace all others."[6] He emphatically dismissed the idea of one single state oil company or monopoly as an "unwise idea that could lead to catastrophic results."

He added that "management of these new companies should remain in the hands of the Venezuelans who are now managing the multinational companies so as to ensure continuity in the process." The figure of the holding company was advanced as the organism for "planning, guidance, and supervision of the four operating companies." He committed himself to naming the board "without resorting to subordinate motives and only looking for those Venezuelans of the highest qualifications and proven patriotism." In his speech Pérez recognized the dangers of mixing the oil industry with a state bureaucratic structure both "inefficient and heavy and with little public spirit." He gave the commission six months to do its job. The commission was composed of thirty-six people representing the cabinet, the political parties, labor unions, universities, professional societies, and economic groups such as Fedecámaras and Pro-Venezuela. However, not one single member of this group was from the ranks of the oil industry. AGROPET did all they could to be included but did not succeed.

The creation of the Nationalization Commission proved to be a wise move on the part of the government. Although the public debate on oil went on, much of it was properly channeled through this task force where all political parties were represented. Public debate became predominantly a confrontation of views between oil technocrats and politicians, largely about the role each sector should play in decision-making and managing of the oil industry. Another feature of the commission was to rob Congress of the spotlight in the discussion about nationalization. The debate in Congress about the two nationalization draft bills of MEP and COPEI lost momentum and became secondary to the work of the commission.

The clash between politicians and technocrats can be illustrated by the incident between Abdón Vivas Terán, a young COPEI leader and congressman and Alberto Quirós, then vice-president of Shell. Vivas Terán said in Congress that the oil companies were juggling their accounting to deposit in the Guaranty Fund less money than they should. This accusation was echoed by Carlos Piñerúa, president of Fedepetrol and MEP labor leader.[7] Alberto Quirós publicly refuted Vivas Terán and asked him either to substantiate his accusations or to apologize. Vivas Terán apologized.

At their ninth national meeting, in Maracaibo in May 1974, the Venezuelan Engineering Society spent much time analyzing the merits of oil nationalization. They decided that an immediate takeover of the oil industry by the state was undesirable, because such a move had to be properly planned. Among the engineers supporting this view were not only members of the private oil companies but also engineers from the Ministry of Mines and Hydrocarbons and from CVP, the state-owned Venezuelan Oil Corporation. The assembly made the following recommendations:[8]

1. The oil industry must continue to pursue commercial objectives.
2. The operational and administrative efficiency of the nationalized industry must be, if anything, higher than at the present time.
3. The nation must have full power of decision in all aspects of industry.
4. Investments must respond for genuine commercial reasons and not to political, geographical, or sectoral interests.
5. Technological transfer must be ensured. Venezuelan technological institutes must be created so as to minimize technological dependence on outside sources.
6. The decision to be adopted should be acceptable to the oil-industry organizations.
7. The resulting scheme must have the backing of oil-company employees.

Also at this time President Nixon sent a special envoy, William Eberle, who declared that the United States had a firm policy on nationalization of American companies abroad, a policy that "respected the sovereignty of countries to dictate their own laws." Exxon's president Clifton Garvin came to meet with President Pérez and stated that Exxon would abide by the decision of the Venezuelan government and that he trusted the government to be fair and equitable. The *Christian Science Monitor* warned the Venezuelan government about taking steps which could set up a chain of similar events in other underdeveloped countries. *Pravda* supported the Venezuelan intentions without reservation, calling it "a decision in the best interests of the people."[9]

The Third Venezuelan Petroleum Congress

In early 1974 the Venezuelan Petroleum Congress was being organized. This was to be the third of a line of events which have always been important in shaping Venezuelan petroleum policy. Held under the auspices of the Venezuelan Society of Petroleum Engineers, these events attracted not only a technical audience, but also a political audience. In fact technical discussions frequently became subordinate to political considerations. The first congress, held in 1962, had been the only one essentially technical and resulted in the publication of a book on the Venezuelan oil industry which even today is one of the most complete and valuable volumes available in the field. The second congress was held in 1970 and had as its central theme a discussion about reversion. This congress became the battleground for the confrontation on that issue between lawyers of the multinational oil companies and the lawyers from the government ministries.

The climate in which the third congress was to be held was charged with emotion and with great animosity toward those who advocated restraint. In spite of the fears of oil technocrats, a considerable number of invitations were sent out to political leaders and congressmen, including representatives of the extreme left. The central theme of the congress was to be "Alternatives for the Future Administration of the Oil Industry." Besides a paper on this theme, several basic papers were prepared to guide discussions—"Marginal Oil Fields," "Technical Services for the Oil Industry," "The New Hydrocarbons Law," "Human Resources and Its Formation," "Development of Research," "Toward an Integral Energy Policy." The organizing committee included Rafael Sandrea, president; Alberto Quirós, vice-president; Arévalo Reyes, secretary general; Gerardo Acosta, secretary of social events; Rubén Chaparro, secretary of information; Romer Boscan, secretary of finance; José Gregorio Paez, general advisor; and myself as secretary of organization. The initial meeting for the congress was held June 8, 1974.

President Pérez gave a speech in which he told the audience that no analysis of alternatives would be required. He said, "Go on and study alternatives, but not for the future of oil, since there is only one: the takeover of the companies by the state, but for the administration (of these companies)."[10] With a single sentence, President Pérez changed the orientation of the discussions to be held from "what to do" to "how to do" what had already been decided at the highest political level. However, Pérez went on to give delegates some leeway. He said, "History has shown that no important step in this world can be taken without risks. This is the moment to be audacious." and added, "I ask from you utmost frankness, that you speak your mind with clarity, that you do not fear any kind of psychological coercion. This is the moment of great national responsibility and it has to be the moment of national frankness." Pérez ended as follows: "Oil in our hands cannot be used to obtain similar returns to those we are presently obtaining . . . but to obtain earnings which we have not, so far, been able to obtain for not having the oil industry totally in our hands."

President Pérez could foresee the clash that would take place between politicians and oil technocrats and could not afford to lose the cooperation of either side. He had to give each side recognition. His call for total frankness was directed to the oil employees because he knew that this group wanted to speak out. At the same time he wanted to let them know that a political decision—that of state ownership of the industry—had been reached and that there was no turning back.

The first paper of the congress was read by its coordinator, Félix Morreo. Dr. Juan P. Pérez Alfonzo, who was not attending the congress, remarked to *Punto,* the Marxist-socialist newspaper of MAS, that Morreo's paper was "garbage" (*bazofia*).[11] The paper was well written and presented four alternatives for the administration of the industry: *state-owned enterprise* directly managed by the state; the *service contract,* by which the state-owned companies would be operated by private contractors; the *mixed company,* in which private capital would have a minority percentage of the shares; and *private administration* under the old concept of concessions, an alternative that had been already rejected at the political level. After Morreo ended his reading, the congress became a veritable circus in which some of the most bizarre specimens of Venezuelan politics were seen and heard attacking those oil industry employees who still dared to question the wisdom of the decision for state ownership. Oil employees fired back, and for two full days the meeting was characterized by extremely harsh language. Politicians such as Arturo Hernández Grisanti from Acción Democrática, Humberto Calderon Berti from COPEI, and Rubén Sader Pérez, former head of CVP, tried to play a moderating role. Others, such as L. Montiel Ortega from URD and Pompeyo Márquez from MAS, were especially acerbic, and even some engineers, including the president of the congress, Rafaél Sandrea, shared their animosity against oil-company employees. So harsh did they become that Román Pacheco Vivas, president of the Venezuelan Engineering Society, felt compelled to say in the closing speech of the congress: "The Board of the Engineering Society will guide future events of this type so as to prevent the repetition of speeches and interventions that tend to constrain the right we all possess of expressing our ideas or which tend to undermine the faith in our institutions."[12] It seemed obvious that Pacheco Vivas, an active member of Acción Democrática, had been asked by President Pérez to insert this paragraph in his speech so as to make amends for the attacks of politicians on oil-industry employees.

The Third Venezuelan Petroleum Congress accorded full support to President Pérez's decisions on oil issues and pledged to provide all the technical advice and cooperation that the country might need to execute those decisions in the most efficient manner. The objectives of political extremism were not achieved. This was due to the courage shown by the oil employees themselves and that of some government officials such as Luis Pláz Bruzual, hydrocarbons director of the Ministry of Mines and Hydrocarbons, who said during the congress: "I am pleased to hear that it is considered desirable to maintain the . . . present oil-industry executives . . . (in charge). Only in this way will we keep the industry free of the long and sad tradition of waste of public resources which is typical of our state enterprises."[13] He added, "There are honest reservations about the best timing and the best way to nationalize." These were brave words to pronounce in such a highly nationalistic atmosphere.

The oil-industry employees became convinced after this experience that they should participate more vigorously in the debate, without fear of being labeled as unpatriotic. Indignation overcame shyness, and, as a result, they became very determined not to let the oil industry lapse—once nationalized—into the same pattern of mediocrity and low efficiency typical of the majority of Venezuelan state-owned enterprises. They turned to AGROPET in increasing numbers, and, by fall 1974 this association had over 2,000 members. AGROPET had chapters in all main oil districts of the country and started a continuous program of participation in the oil debate through the newspapers, radio, television, and through public discussion panels to which they would invite political leaders. The newspapers started receiving the views of oil managers and technicians. The three television stations proved to be quite receptive to their views. Besides the oil technocracy some independent politicians, businessmen, and journalists such as Arturo Uslar Pietri, Rafaél Tudela, Luis Vallenilla, Fernando Báez Duarte, and journalist Carlos Chávez decisively supported a moderate and rational approach to the issue of nationalization.

Panel discussions were held throughout the country. These were open events, the audiences a very mixed bag of technocrats, laborers, housewives, and students. In these panels politicians started having some difficulties in convincing their listeners with their old and worn clichés, whereas the oil employees and managers kept harping on the more concrete issues of exploration, production, financing, manpower, technology, and markets. Politicians started to sound hollow and ignorant of the obstacles to be overcome. At the end of those meetings most people would walk away with the feeling that things were not as simple and clear-cut as the politicians suggested.

In August AGROPET gave their views on nationalization to the president's Nationalization Commission. The presentation was made to a rather small group of attending members and advanced the following recommendations:

1. To establish, as early as possible, a holding company—Petróleos de Venezuela—with a board of eight to nine full-time members. The members should have oil industry experience or, if from outside the industry, solid management experience. No board member should be a career politician.

2. To create three to four integrated companies through a process of simplification of the existing administrative structures.

3. To establish a clear separation between the managerial entity, Petróleos de Venezuela, and the political level. The oil minister should not sit on the board, as was the case with CVP.

4. To create an Institute for Research and Development.

5. To start an urgent exploration program in the continental shelf, in the Orinoco heavy-oil area, and in the areas adjacent to concessions.

The employees of the private oil companies were not the only ones to be invited to present their views to the commission. About eighteen different organizations were invited to express their ideas in a creditable demonstration of their desire to reach a balanced conclusion. Some of the people invited were the highest executives in the industry: A Quirós, vice-president of Shell, and G. Rodriguez Eraso, vice-president of Creole, as well as the board members of the Venezuelan Petroleum Engineering Society and of the Venezuelan Geological Society. Several Venezuelan businessmen, such as Oscar Machado Zuloaga, Hans Neuman, Pedro Tinoco, and José Antonio Mayobre also presented their views.

The Mood of the Country

During September and October 1974 a marketing and operations research company conducted an opinion survey among oil-industry blue-collar workers and white-collar employees as well as other Venezuelans on the nationalization issue. This survey was done on behalf of the president's Nationalization Committee. It showed some interesting results.[14] The great majority of oil-industry employees mentioned "the incapacity of the state to administer their enterprises in an objective, efficient and profitable manner." Even the employees of the state oil company, CVP, shared this perception. According to the sample the lack of administrative skills in state organizations and the constant political interference had turned public administration into an "archaic . . . structure controlled by mediocre, selfish, and corrupted interests of the lowest kind." They added that "such a structure, which permits dishonesty, subsidizes mistakes, . . . where cronyism is rampant . . . cannot guarantee the normal functioning of the oil industry and, much less, its profitability." This judgment represented the perception of a wide range of personnel, from oil workers to oil managers, from private company employees to CVP and ministry employees. According to the survey,

> The most aggressive in their judgment of state ownership and the most pessimistic about the future were oil service company employees. Those who express their views with the most analytic capacity, not devoid of understandable emotion, are those who work for the foreign companies. Those who assume somber and conformist attitudes are the laborers and employees of CVP.

However, there were no signs that a significant number of employees were considering leaving the industry. All seemed to have a strong wish to stay. Their reasons were varied. Some wished to help industry with their experience; others to close the doors of industry to incompetent outsiders.

Some hoped to ensure that company procedures and systems, which had been vital to the success of industry, were kept intact. And some wanted to continue their long careers within a sector they knew well. The survey also indicated that a majority of oil-industry employees agreed with the basic concept of nationalization, but were not very much in agreement with the chosen procedures. In particular they resented that the presidential Nationalization Commission had been almost exclusively structured with politicians. Oil-industry technical and managerial staff had been left out completely. In addition the employees claimed that "the expressions of some of the members of the commission in reference to some oil-industry managers had been offensive and harmful to their reputation by suggesting that they were allies of antinational interests."

The survey was conducted among more than one thousand oil-industry employees, service-company personnel, ministry employees, housewives, students, and independent businessmen. Thirty-eight percent of those interviewed confessed not being clear what the oil nationalization was all about.

The Report of the Nationalization Commission

Considering its large size and dominantly political composition there is no doubt that the presidential Nationalization Commission did a very creditable job. The main group split into five subcommittees: energy, human resources, economic, legal, and operational. Their main achievements were to outline the mechanism to implement nationalization; to recommend a basic organizational model for the state-owned oil industry; to recommend a formula to indemnify the foreign companies to be expropriated; to recommend the treatment to be accorded the oil-industry personnel after nationalization. The work of the subcommittees included very detailed recommendations on their specific fields.

The energy subcommittee recommended that oil production be reduced, that extraction of heavy oils be emphasized, and lighter oils conserved. The refining patterns would have to be changed to allow the processing of increasingly heavier oils, incorporating demetallization and desulfurization plants are required. They also recommended increasing oil and gas reserves by exploration, secondary recovery, the rehabilitation of marginal oil fields, and optimum production, and using gas as a petrochemical input rather than as fuel. As for other resources, the subcommittee urged accelerating the development of hydroelectrical projects to increase oil conservation and evaluating coal resources and other less conventional energy sources. Finally, they recommended establishment of a central energy agency directly reporting to the presidency to coordinate the execution of an integral national energy program.

The human resources subcommittee recommended that the stability of all existing economic and social benefits of oil-industry personnel be guaranteed; that the nationalized oil industry be kept free of political interference in personnel administration; that the organization of the nationalized oil industry be different and independent from public administration; that labor representation be established on the boards of the nationalized oil companies; and that the formation of the new human resources required for the industry be planned and encouraged.

Financial recommendations included ensuring an uninterrupted flow of funds from the nationalized oil industry to the public treasury after nationalization. Indemnification to foreign oil companies was to be based on a time period of payment no less than 7 years nor more than 25; several types of payment bonuses based on different maturity dates; interest rates to be earned by those bonuses no lower than 4 percent nor greater than 6 percent; bonuses largely nontransferable so as to avoid their dumping in the local capital market. The financial subcommittee also recommended the establishment of a five-year cash-flow plan.

The legal group urged that the National Executive be asked to accept a law reserving to the state the industry and marketing of hydrocarbons, and that Congress be encouraged to pass such a law.

The subcommittee on operations suggested a number of interim measures for the period from announcement of nationalization until the actual takeover:

Oversee operations of the oil industry by means of a supervisory committee composed of highly experienced people.

Supply the Ministry of Mines and Hydrocarbons with the required personnel to attend the activities of the supervisory committee.

Conduct an information campaign about the Venezuelan nationalization process among OPEC members and consuming countries.

Evaluate the worldwide tanker situation, so as to be ready in case of interruption of commercial relations with the multinational oil companies.

Reinforce the staff of Venezuelan embassies and consulates in countries that supply oil equipment and materials to the Venezuelan oil industry so as to facilitate the acquisition of such goods after nationalization.

Stimulate the creation of Venezuelan oil-equipment suppliers.

The administrative structure recommended was to be vertically integrated under the coordination of a holding company delegating considerable authority to the operating companies to allow them to accomplish their objectives and plan their investments and operational programs adequately. The oil-industry administration was to be given maximum freedom and flexibility for action within the framework of national planning. It was recommended that the oil-industry administration be allowed to conduct its business according to approved plans, budgets, and programs without the impediment of external controls before the fact. The industry's administration was to be ascribed to the presidency of the republic through a national council of state-owned companies. Direction was to be separated sharply into three levels: political, under the National Executive; managerial, under the holding company (coordinating and advisors staff); and execution, by the operating companies. The holding company was to be staffed with highly experienced managers from both oil and nonoil sectors.

To accomplish the objective of nationalization at a minimum social cost and without loss of efficiency, CVP should continue operating their present areas until it could be assimilated to the new administrative structure. The Ministry of Mines and Hydrocarbons would formulate policy and regulate the activities of the oil industry.

During the first stage, besides the holding company, four operating companies would be created: an international trading company, a maritime transport company, an exploration company, and a technological services company. The second stage would consist of creating as many operating companies as there were concessionaires, maintaining intact their personnel and administrative characteristics. A third stage would emphasize the restructuring of the operating oil companies so as to optimize the use of existing human and physical resources. The possibility of creation of a refining company and of a local marketing company was also to be evaluated during this stage (see figure 4-1).

Comments on the Work of the Commission

The recommendations of the commission were extremely useful to the National Executive in reaching their decisions. The merit of the work of the task force is the greater because of the diversity of interests represented and the limited oil expertise of many of its members, a fact which no doubt put much greater strain on the more experienced members. The work of the highest quality was that of the operational and economic subcommittees. It was well researched and clearly written. These two groups took considerable time to listen to what other people had to say, and many of the ideas and suggestions they advanced were presented to them by outsiders. It is very much to their credit that although many of these ideas ran contrary to their individual preferences they were given very serious consideration and were prominently displayed in the final report.

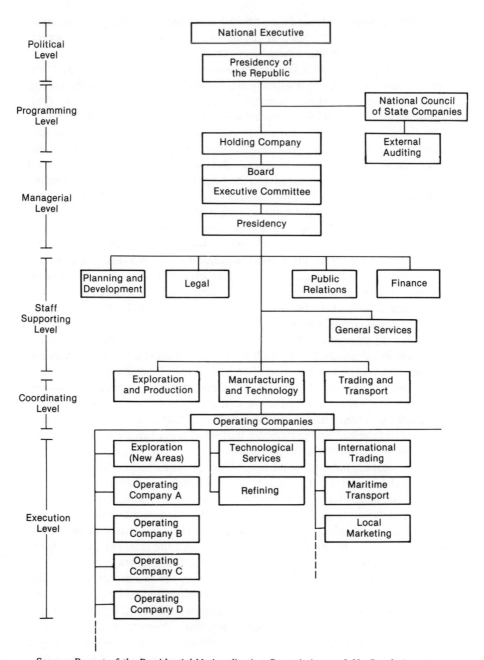

Source: *Report of the Presidential Nationalization Commission,* p. I-53, Graph 4.

Figure 4–1. Proposed Scheme of Organization for Administration of the Nationalized Petroleum Industry

The recommendations of the operational subcommittee had great influence in shaping the structure of the nationalized oil companies. However, the concept of a preventive supervisory committee was not acceptable to oil-industry employees, who felt that intervention in the oil companies was unnecessary. The idea of a supervisory committee was based on the largely incorrect assumption that "Venezuelans do not participate in strategic decision-making, a field which is exclusively reserved to holding companies."[15] The committee was supposed to intervene in the decisions of the industry about oil sales, budgets, organizational changes, promotions, and remuneration plans and long-range planning. But such a supervisory group would have constituted a board parallel to the board of directors of each operating company. Of all the recommendations made by the Nationalization Commission, this was the most unwise since it introduced the probability of a confrontation between government bureaucracy and oil-industry management. President Pérez readily realized this risk and decided not to follow the committee's suggestion. He chose the concept of "observers", who would be assigned to each operating company as witnesses. The observer for Creole was Rafaél Alfonzo Ravard, who would later head the holding company, and the observer for Shell was Luis Plaz Bruzuál, who would also serve on the board of the holding company. This proved to be a beneficial move for all concerned. The observers learned much about management processes in the companies they observed. The companies had a witness but not an interventor to deal with. The state had a formal representative in each company during the months of transition. The Venezuelan government did not want a confrontation and certainly did not need one, either politically or economically, for it rightly felt that most Venezuelans preferred an organized and systematic transition from private to state ownership to a Mexican-type nationalization.

The establishment of an administrative structure to handle the management of the nationalized oil industry was a concept widely accepted. There was also considerable consensus about the principal characteristics such an administrative structure should have. Most of the recommendations of the operational subcommittee were adopted, but there was one very important exception. The subcommittee had recommended the creation of four functionally oriented companies to deal with exploration, trading, transport, and technology and had left open the possibility for two more: refining and local marketing. The idea of functional, nonintegrated companies had been advanced by CVP but did not have the support of most oil industry personnel, who felt that the concept of fully integrated administrative units should be preserved. Functional companies would tend to become cost centers. Furthermore, this concept would have pointed toward the emergence of a single decision-making unit, the holding company. This recommendation did not prosper.

Another recommendation of the subcommittee suggested that the Ministry of Mines and Hydrocarbons should remain, after nationalization, as "the main organism formulating oil policy and controlling petroleum activities." Although the subcommittee added that "it would not be advisable to put in hands of the Ministry of Mines and Hydrocarbons the daily handling of petroleum activities," the suggestion gave support to those in the ministry who wanted to control the oil industry.

The subcommittee identified, quite correctly, some critical factors for a successful nationalization: the attitude of oil-industry personnel; the possibility of political interference in the administration of the oil industry; the ability to sell hydrocarbons successfully in the world markets; the capacity to obtain an adequate technological support to maintain operational efficiency at a high level. These critical factors were taken into account by the government both in their handling of the oil employees and in their negotiations with the multinationals.

Nationalization would affect all workers, technicians, managers, and executives of the industry. To promote cooperativeness special incentives should be created to make them amenable to the event. There should exist clear information, trust, and respect for attitudes and values.

The second critical factor had to do with the danger of politicization after nationalization. The subcommittee correctly perceived the fear of all oil-industry employees that political parties would try to control the administration of state oil companies and that unjustifiable personnel increases would take place. The subcommittee observed that these fears had not been lessened by the promises of President Pérez. Pérez had stated: "To those Venezuelans who have dedicated themselves to study the oil industry and to those who work in the multinational companies, I want to say that they make up a fundamental resource for our country and that no decision in this field will be taken without their previous knowledge and participation."[16] Amelioration of trust, therefore, could come about only by the maintenance of existing norms and procedures regarding personnel development, by the systematic application of the personnel evaluation systems and by the strict consideration of merit as the only basis for advancement. The maintenance of existing administrative structures after nationalization would be indispensable. The subcommittee suggested accepting AGRO-PET's recommendation that experience of no less than 15 years and "a clear professional and technical acceptance" by the oil community be requisites to become a high-level manager in the nationalized oil industry.

In the field of international marketing, the subcommittee recognized the necessity to "contract with the international oil companies . . . replacing these contracts in the medium term by similar contracts with direct clients." In the field of technology, they indicated that the main technological sources existing were the multinational oil companies and recommended the accelerated development of a Venezuelan institute for technological services and research.

In August *El Nacional* published the text of a draft of the nationalization bill that had been prepared by some members of the commission, together with the severe criticism of the draft made by Alfredo Paúl Delfino, president of Fedecámaras and a member of the commission. This publication violated the gentlemen's agreement that no member of the commission would leak to the press information on the work being done. The intention behind the leak was to present Fedecámaras as an enemy of nationalization. This ignited a bitter confrontation between leftists and moderates. The oil workers' union Fedepetrol and Pro-Venezuela attacked Fedecámaras as unpatriotic for their criticism of the draft bill. As days passed, the animosity within the commission increased, endangering the final outcome of its task. By this time, however, most of the essential work had been completed and the task force was finally able to present President Pérez with a final version of their recommendations by late November 1974.

In October 1974 President Ford announced the intention of the United States to diminish oil imports by 1 million barrels a day. He also accused OPEC countries of utilizing oil as a political tool and mentioned the Venezuelan intentions of nationalizing as an example. Ford's speech created an irate reaction in Venezuela. Some political leaders suggested that immediate nationalization of the oil industry would be the best answer to Ford. More responsible leaders such as Arturo Hernández Grisanti, Manuel Egaña, Ramón Escovar Salom, and Rafaél Tudela stated that Ford's speeches could not determine the oil policy of Venezuela.[17]

As 1974 ended, the country was still in the midst of a feverish debate on the oil nationalization issue. The field had clearly split between those who advocated a violent confrontation with the foreign oil companies and those who preferred a nonviolent, negotiated settlement. Among the first group were the political parties of the left, a sector of COPEI, Pro-Venezuela, a group of CVP's employees who felt that CVP should be the sole remaining state oil enterprise after nationalization, and some union leaders. Among the second group were Acción Democrática, Fedecámaras, the oil employees and workers, large sectors of COPEI, and the bulk of public opinion.

The Year of Nationalization, 1975

In January 1975 AGROPET presented President Pérez and his cabinet a series of papers on the issue of nationalization prepared by oil technicians and managers. This was a key event in the debate on nationalization, as it gave the National Executive considerable information on how to structure the industry after nationalization and on the main technical areas needing the attention of the Executive. About four hundred AGROPET members attended the meeting at Miraflores, the presidential palace. The presentation lasted over two hours, during which the following papers were read: "Probable Futures as the Basis for Long-Range Planning," "The Future Corporate Structure of the Petroleum Industry," "Rationalization of Petroleum Operations," "Financial Requirements of the Petroleum Industry," "International Markets," "The Local Market," "The Technological Perspective," "Human Resources for the Petroleum Industry," "The Transition to Nationalization." These papers were the work of dozens of employees. Personal discrepancies were put aside in favor of coherence.

The ideas presented by AGROPET to President Pérez can be summarized as follows:

1. Venezuelan oil policies should be flexible enough to adapt to more than one probable future.
2. The nationalized oil industry should be free from political interference.
3. Operational changes should be gradual, paying extreme attention to the human factor.
4. Financial planning must be carefully made to prevent shortage of funds for the significant investments that would be required.
5. International trading should initially rely on the existing distribution channels owned by multinationals but should progressively be controlled by the state-owned enterprise.
6. Good service should be maintained in the local market.
7. Technological support should be negotiated with multinationals.
8. Human resources should be subject to very professional employment and promotion criteria.

President Pérez was extremely impressed by the determination shown by the oil employees and by the high quality of the papers, especially those dealing with financial matters and with the possible organizational structure of the nationalized oil industry. In a speech following the presentations, President Pérez said,

> I am very impressed by the frankness AGROPET members have shown today in giving their opinions . . . this shows maturity and a strong sense of commitment. . . . They are defeating fear and have entered the national debate with clear opinions and their chosen positions.

He added,

> It is indispensable that those who will have in their hands the responsibility to operate the future state owned companies feel that they are actively participating in the formulation of policy, in the decisions . . . (about) nation-

alization. . . . I understand perfectly what you mean by politicization and your desire that management of the oil industry be conducted free from partisan political influence. . . . I solemnly promise, in the name of the Venezuelan government, that I will be extremely careful to avoid any deviation of that nature in the management of petroleum activities by the state.

The Caracas newspapers gave these presentations prominent coverage and there were numerous favorable reviews and opinions such as the ones by Luis Esteban Rey, Antonio Stempel París, Pascual Venegas Filardo, Marcel Padron, Eloy Porras, Carlos Chávez, Andrés de Chene, *El Nacional, El Universal,* and *La Verdad.* There were also opinions less favorable such as the ones by Jaime Lusinchi, Leonardo Montiel, Radamés Larrazabal, Hugo Pérez La Salvia, Pompeyo Márquez, and Ramon Tenorio Sifontes. Most of the unfavorable comments came from political leaders who felt offended by some of the remarks made by the technocrats. Lusinchi termed the opinion of AGROPET members "inconsiderate and colonized" and argued that the most important decisions regarding the Venezuelan oil industry had been made by politicians and not by technocrats. While President Pérez and many politicians did understand what the oil employees meant, many others chose to react in a very defensive way.

Carlos Piñerúa, president of Fedepetrol, the oil workers' union, also reacted very aggressively. Worried about the white-collar employees taking the limelight in the oil debate, he ordered the distribution of thousands of leaflets accusing AGROPET members of being foreign agents. President Pérez, in an unprecedented move, took to the press and sent a public letter to Piñerúa in which he vigorously defended the oil employees of these attacks. Wrote Pérez,

AGROPET represents, together with the oil workers . . . the most important support we can count on to go ahead with our decision to take over the oil industry. In the meeting that I held with them in Miraflores, they demonstrated capacity, dedication and honesty. . . . I am sure that you share . . . the belief that man has the right to be free from persecution and psychological coercion.[18]

Again President Pérez took the side of oil employees and gave another proof of the immense importance he assigned to protecting this group from harassment.

While this confrontation was going on between oil employees and some labor leaders, another fracas was in the making between Fedecámaras and Pro-Venezuela, the two rivals for the leadership of the Venezuelan entrepreneurial class. Fedecámaras had been for many years the most influential pressure group in the country, although it has never been a particularly well-managed organization. Fedecámaras has always been powerful, its power

being based on the complex and intricate net of family relations and friendships between its members and government bureaucrats. Left-wing political
parties held Fedecámaras in awe. But in spite of a small and distinguished
core of important business leaders, Fedecámaras has been weakened by a
less sensitive class of businessmen, which tends to prosper in the shadow of
government subsidies without risking their own money. Because of this, the
image of Fedecámaras is tarnished in the eyes of many members of the
working class and of the numerous professional middle class.

Pro-Venezuela was also a pressure group but with different characteristics from those of Fedecámaras. It was founded by a nationalistic entrepreneur called Alejandro Hernández, with the original objectives of stimulat-
ing interest for national culture and for the promotion of local industry and
technology. Accordingly Pro-Venezuela traditionally promoted numerous
events such as exhibitions of orchids, books, pottery, or paintings; chess
tournaments and talks about Indian folklore. It also organized frequent
forums about the economy, meetings in which foreign capital and technology almost always emerged as the main villains. Membership was a mixed
bag of orthodox Communists such as the Machado brothers (Gustavo and
Eduardo), conservatives such as José Giacopini, cardinals and archbishops,
Marxist economists such as Domingo Alberto Rangel, and very successful
entrepreneurs like Rafaél Tudela. The overall flavor of the organization,
however, was that of a very nationalistic, almost xenophobic institution.

. Since the presidents of both Fedecámaras and Pro-Venezuela were
members of the presidential Nationalization Commission, it was only a
matter of time before they clashed. Their opportunity came when a draft of
the nationalization bill being prepared by some members of the commission
was given to the press in violation of an existing secrecy agreement, together
with some very critical comments of the draft by Alfredo Paúl Delfino,
Fedecámara's president. Fedecámara's objections centered around two
clauses: one giving total stability to oil workers and another expressly prohibiting foreign capital to participate in the nationalized oil industry. Two
days later Pro-Venezuela's president Reinaldo Cervini made a strong
defense of the draft bill claiming that it was in total accordance with the
Venezuelan constitution.

In the course of the ensuing debate it became apparent that Fedecámaras favored a nontraumatic nationalization, a position shared by the government, Acción Democrática and the oil employees. Pro-Venezuela was
leaning toward a more decisive break with foreign companies. Pro-Venezuela advocated the intervention of the oil companies and objected to the
payment of indemnification. This aggressive posture was influenced by
Domingo Alberto Rangel and some other militant members of the group,
but especially by Reinaldo Cervini, Pro-Venezuela's president.

The confrontation between Fedecámaras and Pro-Venezuela was

understandable. The two groups viewed the issue in different ways. For Fedecámaras oil nationalization was a risky business. If it had to happen, however, it should be cautiously managed and should give local capital the opportunity to participate. For Pro-Venezuela nationalization was an act of independence. In the end the positions of these two organizations canceled each other out, leaving government free to take decisions.

The National Executive Presents Its Proposed Bill to Congress

In early 1975 there were three different drafts of a nationalization bill, two in Congress, and a third one presented to President Pérez by the National-ization Commission. In addition, the National Executive was working on a fourth one. The commission's draft became the one favored by the political opposition. This draft had mainly been the work of MEP's representative Alvaro Silva Calderón and was very similar to MEP's draft. COPEI was now also supporting the commission's draft. Godofredo Gonzalez, acting president of COPEI, said January 15, that COPEI would "oppose any changes in the draft." Octavio Lepage, secretary general of Acción Demo-crática, immediately replied that the commission's draft was "a significant contribution but one which could be improved upon." Fedecámaras insisted that a nationalization could not be done as "an adventure fueled by ideological dogmatism."[19] Juan Pablo Pérez Alfonzo intervened in the debate to say that "my greatest mistake as an oil minister was not to nation-alize the oil industry back in 1945."[20]

In this heated atmosphere Oil Minister Valentín Hernández went to Congress in March to present the National Executive's proposal, a docu-ment similar but not identical to the draft presented to President Pérez by the Nationalization Commission. The draft contained twenty-seven clauses dealing with five main areas: technical (clauses 2–5); administrative (6–11 and 20–22); economic (12, 15–19, 26); legal (13, 14, 25, 27); and labor (8, 23, 24). Clauses 2–4 dealt with the international marketing of oil and stipu-lated that such marketing would be done and controlled by the state and would strive to obtain and conserve "a stable and diversified external mar-ket" trying to obtain "maximum economic returns from exports." Trans-actions should be conducted "preferably with states or state-owned compa-nies." Article 5 included the possibility of having operational agreements in which the state-owned company could contract services on the basis of a tariff—and also, the possibility of association agreements with private enterprises in which the private sector would always have a minority share. These agreements would need the previous approval of Congress.

The *administrative* clauses 6 and 7 defined the nature of the nationalized oil companies, that is, a holding company and a number of operating subsidiaries. Articles 9–11 concerned the supervisory committee. Articles 20–22 dealt with the mechanics of expropriation and with the possibility of utilizing third-party assets (such as drilling rigs). The *economic* clauses dealt with the indemnification to the former concessionaires, forms of payments, possible deductions, and the creation of a guaranty fund.

Legal clauses referred to the alternatives open to the nation in case no negotiated settlement was possible. The *labor* articles had to do with the rights of the employees and type of treatment they would receive when working for the nationalized oil industry. Article 8 defined the oil-industry worker as not being a public employee. The rest of the articles made clear that oil-industry workers would retain all their already acquired benefits and privileges.

In his annual message President Pérez took personal responsibility for article 5, which provided for associations with private entities with previous Congressional approval. Political reactions, both in favor and against this article were very emotional. COPEI denounced the government for "the cramping manner in which it had handled a subject of vital importance." MEP accused the government of "caving in to Fedecámaras's pressures." Carlos Piñerúa, president of Fedepetrol, the oil workers' union, claimed that the government "had changed the rules of the game." The Communist party argued that "the consensus reached by the Nationalization Commission had been violated by the government." The Venezuelan Medical Federation favored "the original and truly nationalistic" version of the law written by the Presidential Commission. The Marxist-socialist parties MAS and MIR also rejected article 5. Juan Pablo Pérez Alfonzo, in a talk given to a group of graduate students of the Universidad Central de Venezuela at Caracas, defined the government's draft bill as unacceptable and suggested that not only the oil industry but all other foreign investments in the country should be taken over by the government. Pro-Venezuela also disagreed with the government's version. In favor of article 5 were Acción Democrática, Fedecámaras, the oil-industry employees, Union Republicana Democrática (URD), and many independent personalities.

Congress opened a final round of analysis of the proposed law, but not before a group of twenty-two congressmen and former members of the presidential Nationalization Commission had introduced to the consideration of the Chamber of Deputies a copy of their version of the law, a move which did not have the backing of MAS. Everybody sensed, by this time, that the debate was ending. Arguments were beginning to sound repetitive, and interest was giving way to boredom.

On August 29, 1975 President Pérez finally announced to the country the putting into effect of the nationalization law. By this time all sectors had said what they had to say. President Pérez declared that the law did not contain any norms or concepts "which would weaken the nationalistic objectives of the state or the fundamental interests of the country." He described the sequence of events which would be taking place after the passing of the law, that is, the creation of the holding company, *Petróleos de Venezuela;* the naming of the supervisory committee to guide the industry during the transition period, August to December 1975; the formulation of an offer for indemnification to the foreign concessionaires based on net book value; and the creation of a guaranty fund, on the basis of which the oil companies would have to deposit a sum equivalent to the 10 percent of their accumulated gross investments to ensure that oil assets would be turned over to the nation in good working order. President Pérez added that Petróleos de Venezuela would be "a company totally free of political interference and would work with the national interest in mind without paying attention to individual interests."

**The Postnationalization Transition,
August-December 1975**

The elements of the plan outlined by President Pérez had already been the object of considerable political and technical analysis. The creation of Petróleos de Venezuela (PDVSA) took place one day after the promulgation of the law by presidential decree, and its first board of directors was named. One week later the supervisory committee was named, and in doing so the government again demonstrated common sense by selecting its members almost entirely from the technical ranks of the Ministry of Mines and Hydrocarbons. Conversations with the foreign oil companies regarding indemnification and the possibility of signing technological and marketing agreements had already been in progress for more than three months. The government negotiating team was led by Oil Minister Valentín Hernández, Julio César Arreaza, and Rafaél Alfonzo Ravard, and had the backing of ministry technical staff such as Arévalo Reyes, Humberto Calderon Berti, Félix Rossi Guerrero, and Alirio Parra. Discussions were not particularly traumatic because oil companies had already accepted being nationalized and could not hope to alter substantially the terms of indemnification. On the other hand they worked to obtain favorable terms in both the technological and the marketing agreements. The political and economic conditions prevailing at the time were such that the oil companies did obtain much of the favorable terms they wanted.

In summary, several factors enabled rapid agreement with the oil companies. The climate was ripe for nationalization. OPEC had emerged as a formidable geopolitical power, and member countries had initiated a series of nationalizations, imposing a trend impossible to reverse. Moreover the

Venezuelan oil industry was almost totally staffed by nationals, and there was considerable domestic political support for the decision to nationalize.

But the transition was not entirely without difficulty. Negotiations with the foreign companies had to be rapid, for delays would have meant the opposition would start complaining about the secrecy of the discussions, weakening the position of the government.

Moreover the international market had contracted after the phenomenal expansion of 1974. The decline in realized prices had caused a 20 percent reduction in Venezuelan export sales as compared to the previous year. A mild winter and an economic recession in the consuming countries had produced a significant reduction in the demand. Multinationals again had alternative sources of oil supply, and inventories were at comfortable levels. The market was once more a buyer's market.

Marketing was largely in the hands of the multinationals. The clients for Venezuelan oil had not been "nationalized," and Venezuela still had to utilize the existing distribution channels of the multinational oil companies.

Technical support for the day-to-day operations and more sophisticated technological advice could still come only from the oil multinational organizations. They had both the capability and the intimate knowledge of the Venezuelan oil industry to guarantee a rapid and efficient response to its needs. The Venezuelan government was negotiating from a position of relative weakness.

The Meaning of the Debate

The national debate preceding nationalization of the oil industry confirmed the trend started in 1958, after the downfall of Pérez Jiménez, by means of which most vital Venezuelan issues had been the object of intense negotiating and bargaining among the different sectors of society. More than a pact of elites, however, the final decision to nationalize the Venezuelan oil industry was the product of a participatory analysis in which the political sectors, the professional middle class, labor, and academia made significant contributions. This explained the freedom from criticism enjoyed by the national oil companies for some years after nationalization and the low degree of partisan politics coming into play when dealing with oil industry issues.

The debate was also important because it forced the oil technocracy to express their opinions publicly. Up to that moment oil-industry technical and managerial staff had been largely passive, saying little outside the highly specialized technical conferences attended by their own kind. When facing an imminent decision about the industry they knew so well, they decided to become an integral part of the debate. In doing so they soon realized how inadequate were their language and demeanor to communicate effectively with the political sector and with the people at large. At the same time, however, they also realized that although the language of politicians might sound more persuasive, it seemed to have a lot less substance. In due time people also started listening to what the technocrats had to say.

The debate clearly allowed for a much more rational outcome. If the decision on nationalization had been made without such ample discussion, it is probable that conflict would have resulted, as the political sector with the exception of Acción Democrática was in favor of a violent event. As more and more groups and individuals contributed their opinions it became evident that the government had support to take a moderate course. And so they did.

Notes

1. Irene Rodriguez Gallad and F. Yanez, *Cronología Ideológica de la Nacionalización Petrolera en Venezuela,* Universidad Central de Venezuela (Caracas 1977), p. 134.

2. For the year 1974 net profits were of about Bs 2.4 billion ($500 million) for the whole industry, a figure more than 10 times lower than MEP's estimate.

3. Files of AGROPET in headquarters of the organization, Caracas. The acronym means Agrupación de Orientación Petrolera.

4. Files of AGROPET, in the headquarters of the organization, Caracas.

5. *El Nacional,* May 18 1974, p. D-1.

6. There were fourteen operating companies at the time of nationalization.

7. Now Piñerúa is a member of Acción Democrática.

8. Documents of the Ninth National Engineering Congress, Maracaibo, 1974.

9. Rodriguez Gallad and Yanez, *Cronología Ideólogica de la Nacionalización Petrolera en Venezuela,* pp. 171, 176.

10. Transactions, Third Venezuelan Petroleum Congress, Caracas, June 1974.

11. *Punto,* June 11, 1974, p. 4. This newspaper is no longer being published.

12. Transactions, Third Venezuelan Petroleum Congress, Caracas, June 1974.

13. Ibid.

14. *El Universal,* November 27, 1974, p. 2-2.

15. This assumption was incorrect because Venezuelan managers and executives were at the highest levels in the boards of operating companies,

and these men did have an important voice in the decision-making. For anyone who knew the industry well, it was unthinkable that a major decision would be taken by Shell without Alberto Quirós playing an important role in it.

16. Page V-75 of the unpublished Report of the Nationalization Commission.

17. Resumen 51, p. 28.

18. *El Nacional,* January 23, 1982, Section D.

19. Rodriguez Gallad and Yanez, *Cronología Ideólogica de la Nacionalización Petrolera en Venezuela,* p. 241.

20. Ibid., p. 246.

5 The Venezuelan Nationalization Model, 1975–76

The Holding Company

The law nationalizing the Venezuelan oil industry reserved to the state all activities relative to the exploration, exploitation, refining, and marketing of hydrocarbons. To accomplish these tasks the law provided for the creation of state-owned companies. Article 6 of the law specified that the National Executive would "create the enterprises considered necessary for the regular and efficient development of said activities" and "assign to one of the enterprises the duties of coordination, supervision and control of the activities to be rendered by the rest of the enterprises."

In order to provide the industry with the funds required for investments, the operating companies would "pay a sum of money to the holding company equivalent to ten percent of the net income originating from their petroleum exports during the preceding month. Article 7 of the law specified that "the enterprises . . . will be governed by the present law . . . . Furthermore, they will be subject to the payment of national taxes and contributions established for hydrocarbon concessions." In this way the state-owned companies would have to contribute to the fisc in an identical manner to that of their predecessors. Article 8 stipulated that "the directors, managers, employees, and workers of the enterprises contemplated under article 6 of the Law . . . will not be considered public employees."

Decree 1123, dated August 30, 1975, was based on article 6 and provided for the creation of a public corporation, domiciled in Caracas, to be called Petróleos de Venezuela (PDVSA). This corporation had as main objectives the planning, coordination, and supervision of activities of all owned subsidiaries as well as the control of these activities to ensure that they were carried out in the most efficient manner.

The initial capital of Petróleos de Venezuela was Bs 2.5 billion, about U.S. $500 million, represented in one hundred shares to the name of the Republic of Venezuela. The holding company in turn was the owner of all shares of the fourteen operating companies. The relationship between Petróleos de Venezuela and these operating companies was meant to be no different from that which had prevailed between the former concessionaires and their respective foreign-based holding companies. The operating companies would look to Petróleos de Venezuela for advice and general guidance, while Petróleos de Venezuela would generally act as a financial holding and as the coordinating agent of oil-industry activities.

The board of Petróleos de Venezuela was made up of nine principal directors and six deputy directors. This group was supported by a staff of coordinators whose mission was that of obtaining and consolidating industrywide information on their respective fields and programming the activity within these fields so that it would be executed by the operating company best qualified to do it. These coordinators had to be high-level managers rather than specialists, and although they had to have an intimate knowledge of their own fields, their main role was one of integrating. There were coordinators for every major field of activity: exploration, production, refining, local markets, international trade, human resources, and planning. The conceptual basis for this type of organization of functional coordinators at holding-company level had been adopted from Shell International. But interestingly enough, the work of these coordinators was to be done through committees made up of members from the operating companies. The extensive use of committees as analytical and evaluating tools was a concept adopted from Exxon. In this manner the basic managerial fabric of Petróleos de Venezuela borrowed from both giants.

The first board, named by President Pérez in August 1975, had as president Rafaél Alfonzo Ravard; as vice-president, Julio César Arreaza; as principal directors, Julio Sosa Rodriquez, José D. Casanova, Edgar Leal, Benito R. Losada, Carlos G. Rangel, Manuél Peñalver, Alirio Parra; and as deputy directors, José Martorano, Luis Plaz Bruzuál, Raúl Henríquez, and me.

Board president Rafaél Alfonzo Ravard was a retired general of the Venezuelan army. He had graduated from the Venezuelan Military Academy, from the Ecole Superieur du Guerre in France, and from the Massachusetts Institute of Technology with a civil engineering degree. He had built a considerable reputation as a manager of public enterprises during his twenty years at the Guayana Development Corporation (CVG), fourteen years as president. General Alfonzo was not only a successful manager but also a skillful survivor in the rapidly changing political environment of Venezuela. Brilliant, highly cultured, rich, and highly respected in political and social circles, Alfonzo Ravard was an obvious choice to head Petróleos de Venezuela. His nomination had the backing of most elements of the economic and political sectors and was viewed with reservation only by the members of the political left, who saw in him a representative of the Venezuelan oligarchy.

Vice-president of the board, Julio César Arreaza, was a lawyer and a member of Acción Democrática, although not a political activist. He had served in the Ministry of Mines and Hydrocarbons under Dr. Juan Pablo

Pérez Alfonzo and had later gone to work for the private sector in the insurance industry. He was a moderate, soft-spoken man in his early fifties who was considered a reliable administrator and who seemed to lack the drive for power that could become an unstabilizing element within the board. His choice was considered to be an indication that the vice-presidency of Petróleos de Venezuela would be the liaison with the government party. Arreaza was considered as a very appropriate choice since he combined political experience with an adequate knowledge of the oil industry.

Julio Sosa Rodriguez was a COPEI sympathizer, a prominent businessman who had important interests in petrochemicals and had been president Caldera's ambassador to the United States. He had a vigorous personality and was particularly interested in the Orinoco heavy-oil deposits, which he felt should be developed exclusively by Venezuela, without any participation of foreign interests.

Director José D. Casanova was a retired petroleum engineer pressed back into service; he had been a manager in the production area for many years. Edgar Leal was a very young bank manager who had only had a brief petroleum career in the Ministry of Mines and Hydrocarbons. His nomination as director had been a surprise, but he proved to be diligent and dedicated. Benito R. Losada was a well-known economist with Acción Democrática affinities. His main wish was to become president of the Central Bank but, while at Petróleos de Venezuela, he did excellent work, although his petroleum experience was very limited. Carlos G. Rangel was a highly respected business leader who had been president of Fedecámaras. Manuel Peñalver was the representative of the labor sector in the board. He was an Acción Democrática labor union leader of considerable experience. Extremely cultured, Peñalver was both articulate and reliable. Alirio Parra was a highly experienced oil consultant who had also served for years in the Ministry of Mines and Hydrocarbons.

Deputy Directors Martorano and Plaz Bruzuál were retired oil executives, both of whom had had long and distinguised careers in the public and the private sector. Raúl Henriquez was Peñalver's stand-in. I had been, up to that moment, a middle manager in the oil industry, and the main reason for my presence on the board was my association with AGROPET, of which I had been president and main motor. There seemed to be little doubt that President Pérez had wished to reward the cooperation received from AGROPET in this manner.

With one exception, then, there were no members of the board chosen from active oil industry ranks. Why was this so? The oil industry certainly had a broad pool of managerial talent, and it would have been logical to assume that the front candidates to staff the board of the holding company should have been the highest ranked oil-industry executives. Up to that moment, however, the oil executives had been working for multinational corporations, and public opinion tended to perceive them as closely associated with those interests. Although the patriotism and honesty of those men were never in doubt, there seemed to be good strategic considerations for keeping them at the operating-company level, at least for the moment. The tacit agreement among all concerned, though, was that by the time the nationalized oil industry settled down to routine, these men would be the logical choice for Petróleos de Venezuela's highest positions.

The first board of directors of Petróleos de Venezuela had to be considered, therefore, as a board of transition. During this very early stage in the life of the nationalized oil industry the relationship between the holding company and the operating companies was that of a loose federation with a rather weak central government providing industry with a very general sense of direction and acting as a welcome cushion between industry and the political world. Resources were very much in the hands of the operating companies.

General Alfonzo recognized this situation clearly. The board members were largely unknown to him and it would take some time to establish a working relationship with them. Meanwhile he thought the best course of action would probably be to let the operating companies run most of the show. He could rely on his friend Guillermo Rodriguez Eraso, president of Lagoven, the largest of the operating companies. Geologist Rodriguez Eraso was a member of Alfonzo's social circle. Alfonzo did not feel equally at east with the other presidents. Most of them were career oil executives who seemed to speak a different language. In particular he seemed to have problems in handling the outspoken and extrovert Alberto Quirós, Maraven's president. He had an aggressive, sometimes abrasive manner and seemed to thrive on conflict. Alfonzo was shy and preferred to deal with crises in a less direct manner.

Alfonzo also realized that he could not maintain a close relationship with every one of the fourteen operating company presidents. He decided, at the outset, that he would meet regularly only with the presidents of the five largest operating companies: Rodriguez Eraso from Lagoven, Quirós from Maraven, Diaz from Meneven, Chacín from CVP, and Quintero from Llanoven. He knew that in a short time these would be the surviving organizations, and in this respect his decision was sensible and pragmatic. The other presidents never really forgave him for this. They felt very strongly that as long as there were fourteen operating companies, each one deserved the attention of the holding company, even if they represented a rather small percentage of production and total manpower. As is often the case, both sides were right, and both sides were wrong. The chief executive officer of a small company cannot expect the same degree of attention from his holding company that the very large operating company would inevitably get, but the chief executive officer of the holding company should have made special efforts to make his small companies feel that they were an important part of the system.

From the start, Petróleos de Venezuela looked and behaved more like a government agency than a private corporation. A visitor walking down the corridors of the eighth floor of the Lagoven building, the headquarters of the new company, could not fail to notice the number of people just standing by—chauffeurs, office boys, secretaries, some sipping coffee, even an occasional shoeshine boy at work. Some secretaries would go from door to door selling tickets for a raffle organized by them or selling clothing to their colleagues.

The working habits of the president did not help to improve this overall air of laxity. He was indifferent to punctuality. His working hours were irregular, and often his meetings would start one or more hours after the agreed time. Even then he would interrupt them or cut them short in favor of a new, unscheduled, event. The meetings of the board or of the executive committee frequently started considerably behind schedule, and the highly paid directors would sit making small talk, waiting for their leader to arrive. And once he arrived, there was no guarantee that the agenda would be adhered to. Most often the general would have a new subject to discuss that might not have anything to do with the list of items in the agenda. He would expand on his chosen subject, sometimes with extreme lucidity, sometimes in a rather rambling fashion, but always in a very fluid manner, so that there was little or no chance for anyone to intervene. The meetings frequently became unproductive or unnecessarily lengthy. As General Alfonzo established this style in his relationship with the board, the board became very passive. With the exception of a few directors, the group listened in silence to the solioquy. This was, of course, a sign of the traditional respect that authority always commands in Venezuela and of sincere admiration for the general's eloquence. As time went by, the board became a group that listened and agreed rather than participated.

The second line of management in Petróleos de Venezuela was made up of coordinators, men in charge of consolidating, at the level of Petróleos de Venezuela, the information on their functions, of digesting this information and programming the activities of their field of expertise in the most efficient manner industrywide. These men came to the holding company from the operating affiliates and generally had two characteristics. First, they were not necessarily the best functional managers the industry could have provided since this would have meant the weakening of the managerial ranks of the operating companies, and second, they still were strongly oriented toward operations since they had been basically operators prior to coming into PDVSA. As a result, they saw their new jobs as line rather than staff positions. They tended to become line managers in their interaction with their counterparts at the operating company level. The exaggerated interest of some of the coordinators in the day-to-day activities of the operating companies became one of the first and most obvious distortions of the relationship between PDVSA and its affiliates. One clear example of this distortion can be found in the early tendencies shown in the coordination of exploration. This group defined its job as one of "obtaining, analyzing, interpreting, and evaluating all information on the prospects for hydrocarbons in the country, looking for these prospects and estimating the possible reserves to be found." To do this, the coordination estimated that the exploration staff of Petróleos de Venezuela should be of some sixty persons by 1978–79. The board seemed to agree to this plan since the president of Petróleos de Venezuela favored the concept of a strong holding company. This was interpreted by the members of the board to mean a *quantitatively* strong holding company instead of a *qualitatively* strong holding company. The application of the concept of a quantitatively strong holding company by all coordinations of PDVSA would have led to the creation of a very large organization, deeply concerned with day-to-day operational activities instead of adhering to its original objectives of coordination, control, and strategic guidance. Fortunately some other oil-industry executives, notably in the operating companies, did not see things the same way. Although the views of these executives introduced a reasonable balance in the philosophy of growth of PDVSA, the tendency to centralize operations never left the minds of several of the most influential members of the board and would constitute a major source of disagreement for some time to come.

Main Characteristics of the Nationalization Model

The main features of the nationalization of the Venezuelan oil industry added up to a unique administrative model. This model was the result of many minds coming together during the national debate of 1974 and 1975. It also had the benefit of the experience the government had obtained when nationalizing the iron industry, in 1974. In the case of the iron-ore concessions, the government of Carlos Andrés Pérez made some decisions which would create important precedents for the nationalization of the oil industry. One was the payment of compensation on the basis of net book value and the use of "promissory notes at 7% interest."[1] The other was the signing of technical assistance and marketing contracts with U.S. Steel and Bethlehem Steel. As in the case of the oil nationalization, the main objective of these agreements was to ensure an uninterrupted flow of operations. As Sigmund observes, "The iron nationalization provided a useful model for the subsequent and much more important oil nationalization."[2] One basic element of the oil nationalization model was the national industry's relationship with the former concessionaires, which depended on negotiations leading to a settlement, compensation based on net book value and payment in bonds, technical assistance agreements, commercial agreements, and establishment of a guaranty fund. The basic organizational structure of the model was of course the holding company and multiple, integrated affiliates, as well as a research and development institute.

Managerial philosophy included commitment to keeping the large organizations intact, professional management, "meritocracy," or the merit system in career development, and self financing. Basic strategies stressed were operational continuity, modernization of plant and equipment, generation of new reserves, and careful manpower planning.

Relationship with the Former Concessionaires

Negotiations. The nationalization of the Venezuelan oil industry came about after intense negotiations with the foreign oil companies. The foreign concerns were amenable to a takeover by the state since the control of the industry had already been essentially in the hands of the government since 1972. Negotiations were thus directed toward establishing the amount and type of compensation. The negotiated nature of the takeover was clearly shown by the inclusion of articles 12 and 13 in the nationalization law. Article 12 reads as follows:

> Within 45 calendar days following the promulgation of this law, the National Executive, through the minister of Mines and Hydrocarbons, will present to the concessionaires a formal offer of compensation for all the rights they have with respect to the assets subject to the concessions to which they have title, compensations to be calculated pursuant to the provisions of article 15 of the present law and to be paid according to articles 16 and 17 of same. The concessionaire will answer said offer within the next 15 calendar days after the receipt of the corresponding communication to be made by the National Executive. If an agreement is reached, a written document will be signed to that effect between the attorney general of the republic, according to instructions given to him by the National Executive through the minister of Mines and Hydrocarbons, and the corresponding concessionaire, effective the date of extinguishment of the concessions as provided for under article 1 of the present law. The National Executive, acting through the minister of Mines and Hydrocarbons shall immediately submit said written document to the Houses of Congress, acting in joint session, for their consideration and approval, it being understood that said Houses must come to a decision within the shortest possible term, which in no case shall exceed 30 calendar days following the date of receipt.

Article 13 provided an expropriation mechanism in case agreement with the concessionaires was not reached. Article 14 stipulated the provisions for the previous occupation of the assets subject to expropriation. This would be done by means of an order issued by the Supreme Court. As will become evident, the government took careful steps both to reach a negotiated settlement with the concessionaires and, if negotiations failed, to expropriate in an impeccably legal fashion.

Compensation. Article 15 of the law dealt in great detail with the mechanisms of compensation. The essential point was that "the amount of compensation of the expropriated assets cannot be higher than the net value covering properties, plants, and equipment, its being understood as its value of acquisition less the accumulated amount of depreciation and amortization as to the date of the petition of expropriation and as per books utilized by the concessionaire for income-tax purposes."

From this amount, the following deductions were to be made:

The value of assets which were subject to reversion according to that law and for which official communications ordering their return to the nation were still pending

The value of petroleum extracted by the expropriated concessionaires beyond the limits of their assigned reservoirs

The amount of the social benefits and other employee rights as defined by article 23 of the nationalization law

The amounts that the respective concessionaires were indebted to the National Treasury and other public organisms

In addition, all litigations in progress involving the concessionaires would continue and, if they were found liable, deductions would be made from the compensation or from the guaranty fund contemplated under article 19 of the law. Payment of compensation could be deferred for up to 10 years and could be made in bonds of the public debt, which would earn an interest no greater than 6 percent per annum.

The actual compensation came up to $1,012,571,901.67.[3] Net compensation is still not known, because tax claims dating back to the early 1970s are still pending. This amount of compensation was several times lower than the replacement value of the assets involved in the takeover, a value that has been estimated at no less than $7 billion in 1977 prices.[4]

Technical Assistance Agreements. The analysis made by the Venezuelan Petroleum Institute (INVEPET) and the recommendations of oil-industry employees had left little doubt that the operational continuity of the oil industry could not be ensured unless there was a substantial inflow of technical and technological assistance from outside sources. The INVEPET report contained the results of a survey conducted in the Venezuelan refineries to identify the necessity for outside help and the best sources of the required help.[5] Of a total of 205 items of technical support received by these refineries from outside sources at the time of nationalization, 33 percent were considered indispensable, 39 percent highly desirable, and 28 percent desirable; none were considered to be unnecessary. Eighty-three percent of this support was contributed by the multinational oil companies. Ninety-seven percent of the refinery managers and technicians who were consulted said that to replace this external support by internally generated technology would take 5 years or more.[6]

The report also concluded that the operating oil companies were receiving efficient technological support at the time of nationalization. To interrupt this inflow would diminish efficiency. It was desirable therefore to negotiate technological agreements with the concessionaires.[7] It was quite clear, therefore, that the Venezuelan industry required technical support and that the obvious source was the companies which had been in the country for many years. This was easy to understand. Most of the plants and equipment of the Venezuelan oil industry had been built or installed with the intimate support of those companies, mostly to specifications supplied by them and similar to the ones these companies owned in other areas of the world. Exploration and production drilling, secondary recovery techniques, and the myriad of activities in the industry all had benefited for many years of a process of technical transfer. Of equal value were the personal connections. The hundreds of high-level oil specialists, managers, and executives now working at the central offices of the multinationals had almost invariably spent long years in Venezuela and were conversant with local conditions, usually knew the language and many of the older people in the industry. Such connections doubtless allowed for a rapid and efficient response to the Venezuelan requests for technical assistance. To bring in totally new technical advisers at that moment would have been a questionable managerial decision.

The government decided therefore to negotiate technical assistance agreements with the multinationals. The nature of this assistance was to be highly diversified. It included the assigning of specialized personnel to work as advisors within the ranks of the national operating companies, the access to new technology, the support to solve the day-to-day operational problems, visits of technical audit teams, the use of computer programs, the receiving of technical documentation generated by the multinationals, the training of nationals; in fact, a global mechanism of technical support. This support would be given in all fields of activity of the industry, including exploration, production, refining, local marketing, administration, training, computing, training on the job of employees, environmental protection, systems, technical documentation, materials acquisition abroad.

Through the agreements about three-hundred foreign specialists remained in the nationalized oil industry, almost all of them acting as advisory staff, although a few of them had line authority. Contracts were signed with all of the previous concessionaires, mostly for four years, although some agreements with smaller companies were only signed for 2 years. One company, Atlantic Richfield, objected to this mechanism and chose not to sign an agreement with the nationalized oil industry.

Payments under these contracts were tied to production and refining volumes and averaged about 14 (U.S.) cents per barrel. Some of the larger companies such as Shell and Exxon obtained a fee of about 15 cents per barrel (after tax). For a production of about 2.2 million barrels a day, this meant that some Bs 700 million in fees (about $160 million) would be paid to the multinationals for technical assistance. This represented less than 2 percent of the sales value of the produced hydrocarbons.

The signing of the agreements brought about a violent reaction from the political left and even from some sectors of the moderate, Christian Democrat COPEI party. These groups claimed that the technical assistance agreements were no more than a disguised form of compensation and that, through them, the multinationals would now get more profits than before nationalization without running any risks and without having to bring capital into the country.

Although the allegations were not altogether unfounded, these political groups failed to see the agreements in a more global context. The agreements had three positive characteristics: First, they could guarantee the uninterrupted operations of the Venezuelan oil industry. Second, the fees represented a very small percentage of the total value of the oil production. It was estimated that, without technical assistance, the productivity of the industry would have decreased by 10–20 percent. This would have meant a yearly loss of $850–1,500 million or 5–10 times as much as was being paid in fees. The third advantage of the agreements was that the fees paid by Venezuela in technical assistance were among the lowest of all oil-producing countries. Iran was paying 15.4 (U.S.) cents per barrel, Saudi Arabia and Qatar about 15 cents per barrel and Venezuela about 14 cents per barrel. Only Kuwait seemed to be paying less, but their production was much easier to obtain than the older and very complex Venezuelan production.

In addition the critics of the agreements failed to recognize the very difficult conditions under which the Venezuelan government had had to negotiate the agreements. Among the difficulties faced by Venezuela at the time of the negotiations were decreased demand due to a very weak oil market and low tanker rates, which made African and Middle Eastern crude oils very competitive. The complexity and old age of the Venezuelan hydrocarbon reservoirs made production a difficult task; the task was complicated by the number and diversity of refining installations and the organizational complexity of the industry. These and other factors put the negotiating team composed of ministry officials such as Arévalo Reyes, Humberto Calderón Berti, Félix Rossi Guerrero, and Félix Morreo at a disadvantage vis-à-vis the multinationals.[8] The main objective of the agreements was as defined by Oil Minister Valentín Hernández, "to obtain the same degree of technological assistance that the concessionaires were receiving up to December 1975."[9] In spite of all the difficulties, the agreements finally reached were essentially beneficial to the country and the main objective, as outlined by Hernández, was achieved.

Commercial Agreements. At the time of nationalization, world oil markets were weak, especially the markets for heavier crude oils of high sulfur content such as made up about 40 percent of Venezuelan exports. Venezuelan production was 2.2 million barrels per day, but during the early stages of the negotiations, the multinational concessionaires had indicated that the Venezuelan crude oils were overpriced. The government stood firm in its price requests even if that meant the selling of much lower volumes. As a result the first postnationalization sales agreements were closed on a total of only about 1.4 million barrels per day, 600,000 barrels short of the Venezuelan expectations. The agreements were to last for two years, but prices would be revised every three months.

The commercial agreements negotiated by the Ministry of Mines and Hydrocarbons and the multinationals also ran into violent opposition from diverse political sectors. Former Oil Minister Juan Pablo Pérez Alfonzo argued that Venezuela should have sold their oil through a mechanism of international bidding. Pérez Alfonzo had already experimented with this system in the 1940s when the government had tried to sell some of its royalty oil. That experiment was unsuccessful even though the market was strong. There was little chance that such a system would work any better under much weaker conditions of world demand as the ones prevailing in 1976.

The power of negotiation was not in the hands of the Venezuelan team. The Venezuelan oil industry, with the exception of Maraven (the former Shell), had not established international trading organizations, and as a result practically all of the Venezuelan oil exports were sold through the trading channels of the multinationals. This almost absolute dependence on the multinationals for the selling of their oil put the Venezuelan nationalized industry in a situation of extreme negotiating weakness. All Venezuela hoped to achieve was the execution of reasonable agreements as well as initiating a vigorous process of diversification to minimize dependence on the multinationals in the short term. The industry was so successful in doing this that, by 1979, almost 70 percent of Venezuelan exports were being marketed outside the multinationals' trade systems.

The Guaranty Fund. Article 19 of the nationalization law established that

> For the purpose of guaranteeing the fulfillment of any of the obligations imposed by the present law . . . the guaranty fund, provided for in the law concerning properties subject to reversion in hydrocarbon concessions is modified. . . .

as follows:

> The concessionaires must deposit, in one payment, the necessary amounts equal to a sum equivalent to the 10 percent of their accumulated gross investment accepted for purposes of income tax.

Article 20 added that

> The National Executive will effectuate the fiscalizations and inspections in order to ensure the physical existence of the assets expropriated by the nation, as well as their state of conservation and maintenance, within a period of time not to exceed three years from the time said assets are received.

By means of these two articles, the concessionaires were not only obliged to deposit important amounts of money into Venezuelan banks but to wait a considerable amount of time to get their money back. In fact as late as 1982 the companies were still trying to do that since tax claims still pending and other legal or political complications prevented them from doing so. This became a major source of friction between multinationals and the Venezuelan government.

The Basic Organizational Structure

Reporting to the holding company Petróleos de Venezuela were fourteen operating companies, the inheritors of the former foreign organizations. These new companies were organizationally identical to the previous concessionaires. Only the name and the ownership had changed. They had kept their administrative structures intact, were managed by the same teams (with the exception of the foreign executives who were gone), and in general fully retained their own identity and strong esprit de corps. Some of these companies such as Lagoven (formerly Exxon), Maraven (Shell) and Meneven (Gulf) were very large and, between them, accounted for about 80 percent of total oil production and manpower. Some others were very small, such as Vistaven (formerly Mito Juan) and Taloven (Talon), which together accounted for less than 1 percent of the production. It was obvious that in time something had to be done about discrepancies between the large and the small.

The companies had important differences in management style, managerial talent, and technical expertise. Some were more efficient than others, although their operations were different enough to make a comparison between them somewhat difficult. It might be interesting to give a summarized profile of the most important of these organizations since this might help to understand some of the events which took place later on.

Lagoven. At the time of nationalization this company (formerly Exxon) had a production of about 1 million barrels per day (40 percent of the country's total) and was refining some 320,000 barrels per day. It had about 8,000 employees and operated in more than 1 million acres of territory in both western and eastern Venezuela. Its proven oil reserves were about 7.3 billion barrels, about half of the Venezuelan total for 1975. Lagoven operated the largest refinery in the country, Amuay, in the Paraguaná Peninsula. It had a very experienced board of directors, including geologists Guillermo Rodriguez Eraso, the president, Ernesto Sugar and Nicanor García and solid administrators such as Jack Tarbes. Lagoven was highly disciplined, a trait inherited from its predecessor Creole Petroleum Corporation, with a strongly ingrained tendency to make crisp and timely decisions. The board members acted as corporate directors with a very remote functional authority. They were only liaison between the board and the department heads, who were the real day-to-day managers of the company. Much of the high-level decision-making was done by committees composed both of directors and managers, groups which met regularly to consider carefully structured agendas.

Lagoven's strengths were in light-oil production and in their high standards of maintenance and generally superb organization. Their weak points had to do with international trading, corporate planning, local human resources in the finance sector, and with their traditional secrecy, which, under the atmosphere of nationalization, had rapidly turned into a liability.

Maraven. In 1975 this company (formerly Shell) was producing about 540,000 barrels per day (25 percent of total production) and refining about 300,000 barrels per day. It had about 6,200 employees and its operations were restricted to western Venezuela. Proven reserves were 3.5 billion barrels. Maraven operated the second largest and most sophisticated refinery in Venezuela, Cardon, also on the Paraguaná Peninsula. Its board of directors included Alberto Quirós, president, and Carlos Castillo, vice-president—the two best international marketing experts in the industry—as well as very experienced geologists and engineers such as J.R. Domínguez and Ramón Cornieles. Maraven had a very open, participatory management style in which conflict was openly dealt with. More so than Lagoven, Maraven had been run as a very autonomous unit in concessionary times and had been a regional center controlling a large group of smaller companies in the Caribbean and South America. As a result Maraven had created an important group of strong functional managers. Contrary to Lagoven, Maraven made little use of committees and gave their directors much more functional, line authority. By the time of nationalization, Maraven decided to move away from an almost purely functional board toward a more corporate one.

Maraven was very strong in heavy-oil production and upgrading since it had operated the heavy-oil fields of Lagunillas, Bachaquero, and Tia Juana for many years. It also had much expertise in international marketing and finance, and in general, more management depth than any other of the Petróleos de Venezuela affiliates. Weak areas were in deep drilling operations, manpower control, and personnel planning and training.

Meneven. The successor to Gulf had a production of about 400,000 barrels per day and refined less than 100,000 barrels per day. It had about 3,800 employees and proven reserves of close to 2 billion barrels. Meneven operated the Puerto La Cruz refinery in eastern Venezuela. Its board of directors was weak since Bernardo Diaz, president, and Francisco Prieto, director, were its only members with considerable management experience.

Never an autonomous corporation, Meneven had been a strong operating, producing, center for Gulf. All major decisions were made by Gulf in the United States, with the Venezuelan operators mostly receiving yearly production targets. As a result Meneven had no corporate managers, though it did have good operators. This weakness became very apparent at the time of nationalization because entire areas of special strategic importance such as finance were almost exclusively staffed by expatriates of other countries. When the expatriates had to be removed, their replacements generally lacked managerial experience.

Meneven was a very good production organization and had had considerable experience in Eastern Venezuela. The Puerto La Cruz refinery was very well run with an emphasis on cost control. The main problem of Meneven was the absence of good managers, even at the highest levels of the organization.

Llanoven. This company (formerly Mobil) produced about 100,000 barrels per day (5 percent of total production) and refined 150,000 barrels per day at the El Palito refinery, located near Puerto Cabello. It had about 1,200 employees and operations all over the country; proven reserves were of only 200 million barrels. The organization had been tightly run for many years, more as a foreign division than as a truly integrated overseas subsidiary. It seemed to be a candidate for consolidation with other medium-size organizations.

CVP. This organization had a production of about 110,000 barrels per day, practically no refining capacity, and proven reserves of about 900 million barrels. It had about 3,000 employees.

CVP's problems were, an acute lack of managerial talent, as in the case of Meneven, and also a lack of efficient administrative systems and procedures. In addition CVP was starting to suffer from significant political interference. The board of directors was rather weak and seemed content with keeping the organization small. After its failed bid to become the only state-owned oil company, the drive of the organization had practically disappeared.

Recall that the basic structure of the nationalized petroleum sector included an Institute for Research and Development. The origin of this institute had been at INVEPET (the Venezuelan Petroleum Institute), an appendix of the Ministry of Mines and Hydrocarbons created in 1974 to start gathering information on technology. In May 1976 INVEPET became an affiliate of Petróleos de Venezuela under the name INTEVEP (the Venezuelan Technological Petroleum Institute). Its main objectives were to serve as the research and development institute for the Venezuelan petroleum and petrochemical industries and to minimize the technological dependence of those industries on foreign sources. The man in charge of the Institute in 1976 was Humberto Calderón Berti, who would later become oil minister in President Luis Herrera's cabinet. Although the president of the institute was José Martorano, an appointee of Petróleos de Venezuela, the executive in charge was Calderón Berti. When INTEVEP became an affiliate of Petróleos de Venezuela, the original intention of the executive committee had been to remove Calderón Berti from the institute on the grounds that he was more of a politician than a stable technocrat. But since he had done an excellent job during the early stages of the organization, a compromise was reached to retain him as the executive vice-president and to name José Martorano, an aged and respected member of PDVSA's board, as INTEVEP's symbolic president.

Managerial Philosophy

The Venezuelan oil industry was nationalized as a going concern. It had more than five decades of successful management traditions and its different organizations possessed very distinctive organizational cultures. Shell, Creole, Mobil, and the other concessionaires had created through the years groups of men and women fiercely loyal to their organizations and with a strong esprit de corps. It seemed logical to keep these organizations largely intact, each with its own leaders, staff, systems, and formal and informal organizational cultures, since this would allow for a healthy competition, rather than to use the concept of a state monopoly, so often unsuccessful in other nationalizations.

The leaders of these organizations were, almost without exception, proven professional managers. These men had no shares in the companies they managed. They were extremely well paid to make decisions which would give their companies optimum benefits. To replace these men at the helm of their organizatios in favor of personal or political appointees was thought to be unwise. They were retained and given assurances that eventually they would have access to the highest positions of the industry.

These assurances were clearly stated, in different circumstances, by President Carlos Andrés Pérez and by the most prominent political leaders of the country. The merit system, they claimed, was the only one which should be used in establishing promotions within the industry. If anyone had said in 1976 that the board of Petróleos de Venezuela would someday be chosen on the basis of personal favoritism, he would not have been believed.

The other solid ingredient of the managerial philosophy of the nationalized oil industry was self-financing. For a state-owned enterprise to be efficient it had to be self-financed was the consensus. It could not risk going through the political system to obtain the capital required for investments because in a strongly politicized environment, such as the Venezuelan, this could probably mean long delays, distortion of original objectives or, even worse, opening the doors to large-scale corruption. Self-financing was achieved through the retention in the financial system of Petróleos de Venezuela of 10 percent of the net value of industry exports and of the net profits of the operating companies.

Basic Strategies. In 1976 the most basic of strategies for the nationalized oil industry consisted of keeping operations going. The country needed the oil payments day by day and could not afford an interruption in the operations of the industry. The need for continuous activity prevailed over most other considerations and explains the signing of the technological and marketing agreements with the multinationals, even if some of their terms were not the most favorable for the country.

The second short-term objective of the Venezuelan national oil industry was the modernization of plant and equipment. These facilities were partially obsolete at the time of nationalization, and replacement of substantial portions was urgently needed. This posed a formidable logistic problem, since these needs coincided with a worldwide increase in petroleum activities brought about by the higher prices of hydrocarbons in the international market.

Equally urgent was the need to generate new oil reserves to replace the diminishing proven reserves of light and medium oils. Obviously a great effort would have to be made in offshore drilling to try to find new oil and gas fields, and this effort would require heavy investments and the rapid buildup of an exploration task force.

Manpower was the other area meriting immediate attention. Once the industry had been nationalized and long-range planning was again feasible, there was a clear need to create the manpower which would be needed for an expanding oil industry engaged in very large projects such as the development of the heavy-oil deposits of the Orinoco area and the construction of new refineries and secondary recovery projects. The extractive stage should clearly end in favor of a more balanced industrial expansion, which would require numbers of new qualified personnel.

Notes

1. P. Sigmund, *Multinationals in Latin America* (Madison: University of Wisconsin Press, 1980), p. 238.

2. Ibid., p. 239.

3. Ibid., p. 243.

4. *Resumen* 167, January 16, 1977, p. 30.

5. INVEPET, "Diagnóstico sobre Transferencia Tecnológica de la Industria Petrolera," Caracas, 1975.

6. Ibid., p. III.22.

7. Ibid., p. II.31.

8. *Resumen* 136, June 13, 1976, p. 24.

9. Ibid., p. 23.

Part II
Performance

6

The Industry in January 1976

The prevailing mode in the nationalized Venezuelan oil industry was one of contraction. Being faced with reversion in 1983 oil companies largely refrained from investing and concentrated on producing oil at the lowest possible cost. In the previous 15 years, investment levels had rarely exceeded $300 million per year, an amount clearly inadequate to keep such a complex industry modern and vigorous. The industry had low exploration levels, a poor record of discoveries, and a high rate of production—a combination strongly suggestive of a policy of liquidation. Oil production had increased very rapidly in the late 1950s, going over 1 billion barrels per year in 1959 and staying at some 1.3 billion barrels per year throughout the 1960s. Oil reserves on the other hand had started to decline in 1965, reaching 13.8 billion barrels in 1973. During that year booming oil prices had caused a redefinition of hitherto uneconomic heavy oil reservoirs, on the basis of which about 5 billion barrels of heavy oils from the Boscán, Bachaquero, and Lagunillas fields were added to the category of proven reserves, increasing the level of these reserves to about 18.5 billion barrels. Discoveries represented only 15–20 percent of the yearly additions to proven reserves, however, with the bulk coming from extensions and revisions, mostly through a change in the parameters used in the mathematical calculations of reserves.

Exploration

The level of exploration activity was low for several reasons: the no-more-oil-concessions policy formulated by Oil Minister Juan Pablo Pérez Alfonzo during the Betancourt regime of 1959–1964; the increase in the complementary oil taxes on excess profits from 26 to 45 percent in 1958, and, above all, the creation in 1960 of the Corporación Venezolana del Petróleo—CVP—the state oil company, which obtained the monopoly to explore in acreage outside existing concessions.

The no-more-oil-concessions policy, combined with the announced plans for reversion by 1983, made foreign oil companies extremely reluctant to invest in exploration ventures, in which returns at best start to be obtained only after 8 to 10 years. The fact that they could no longer explore outside their relatively well-known concessions made exploration investments even less attractive.

Up to 1960 exploration activity still was adequate, because the companies had been evaluating the concessions received in 1956 and 1957 from the Pérez Jiménez government. As these evaluation programs ended, exploration came almost completely to a halt. Exploration wells decreased from about 150 in 1958 to fewer than 70 in 1961, and geophysical and geological activity dropped to its lowest level since 1933.[1]

The reduction in exploration activity brought about a decrease in experienced geological and geophysical personnel. During the 1950s there were no less than 800 geologists and geophysicists working in the Venezuelan oil industry. By 1961 fewer than 100 remained, and in 1976 fewer than 40 were still engaged in exploration. This would be one of the most severe constraints the industry would have to face after nationalization.

Reserves and Production

Proven oil reserves stood at 18.5 billion barrels in 1975. This level was relatively comfortable since it allowed for a reserve to production ratio of about 21 years, the highest in a decade. However, the composition of the oils gave cause for concern. About 60 percent of that figure was heavy oil with a relatively high sulfur content and low commercial value in the world markets. Production, in contrast, was made up of almost 80 percent light and medium oils (about 23° API) which meant that the reserves of these higher gravity oils were being depleted at a disproportionately rapid pace. This emphasized the need for an aggressive exploration campaign designed to find light oils as well as for a revision of the existing production and refining patterns to make possible the processing of much more of the heavy oils.

The Complex Infrastructure

In contrast with Middle Eastern oil-producing countries, in Venezuela the production and refining patterns were extremely complex. By 1975 Venezuelan production came from more than 9,000 individual reservoirs. The crude oils ranged from 8° API, heavier than water, to 50° API, practically natural gasolines. There were more than 250 oil fields dispersed throughout the country. The oils produced in these fields were collected into about 85 streams, which in turn gave rise to 45 commercial segregations sold under the name of the field or area from which they originated. Oil and gas were transported via a network of pipelines of more than 10,000 kilometers.

In 1975 production came from about 7,000 active wells, although during that year about 14,000 wells were closed, many of which were potentially productive. In fact by 1976 the number of active wells went dramatically upward, to almost 12,000. But this increase in the number of active wells was not reflected in a corresponding increase of production. The explanation for this is not entirely clear, although it must be at least partially sought in the very high rates of decline of the majority of the old reservoirs. There is also the possibility that much production potential was permanently lost when wells were closed, in 1975, to conform to instructions by the Ministry of Mines and Hydrocarbons. This could well be the case because in 1974 there were 12,253 producing wells in the country and the average production during that year was almost 3 million barrels per day. In 1975 the production had dropped to 2.34 million barrels per day, a loss of about 600,000 barrels per day which were never recovered.

Production potential in the Venezuelan mature oil industry had a high rate of decline, on the order of 25 percent per year. This meant that production would drop dramatically within a very short time unless intense drilling and repair activity was conducted. This explained the ever higher costs of maintaining oil production at the same level year after year.

The Refining Pattern

The refining sector also presented a major challenge. Although the Venezuelan refining capacity was the largest of all OPEC countries, the yield of products was becoming increasingly undesirable. These refineries had been built by the multinational companies with the U.S. residual fuel market in mind, a strategy that had served its purpose well for many years. In 1975, however, the pattern of consumption in the United States, especially on the East Coast, was changing rapidly. Environmental considerations and conservation policies, including an effort to utilize coal and other domestic energy sources, started to weaken the demand for Venezuelan residual fuel oils, especially those with a higher sulfur content (2.5 percent). At the same time the Venezuelan domestic market for gasolines and middle distillates had grown very rapidly, at more than 11 percent per year. The necessity for higher gasoline outputs and the increasing problems in marketing the barrels of fuel oils produced with every barrel of gasoline, made imperative a drastic revision, in the short term, of the national refining pattern. This would be a major undertaking, requiring not only substantial investments but also a considerable amount of project management expertise.

Trade

Venezuelan exports in 1974 averaged about 2.8 million barrels per day, but already in 1975 the demand in the international market had softened considerably, to the point that exports by January 1975 had fallen to no more than 1.5 million barrels per day. Most of the oil could only move through the multinational oil companies. Alternative outlets had to be found. In 1975 only Maraven, the former Compañia Shell de Venezuela, had an international marketing division. This company had been marketing its own crudes and products for more than 12 years and had formed an important group of well-trained middlemen as traders and well-established connections with final clients. In the U.S. East Coast market the company used Asiatic Petroleum Company as their commercial representatives, but in most other areas it sold directly to clients. This was not true of Lagoven (the former Exxon) or Meneven (the former Gulf), which still depended entirely on the former holding companies to sell their hydrocarbons.

The Local Market

Local consumption in 1975 was almost 80 million barrels of products, that is, about 200,000 barrels per day, half of which was gasoline. Prior to nationalization President Carlos Andrés Pérez had announced that service stations would be sold to third parties; as a result only 10 percent of service stations were owned by the industry at the time of nationalization. Their number was already clearly insufficient to provide for good and fast service to motorists. Something had to be done in this area rather quickly since service stations tended to be seen by public opinion as a symbol, good or bad, of nationalization. Dirty, ill-run service stations would surely transmit a message of inefficiency to the public, who would extrapolate that impression to all other activities of the nationalized oil industry.

The Condition of Plant and Equipment

A prominent source of concern for the industry in 1975 was the working condition of plant and equipment. The philosophy of contraction prevailing in the industry for several years had meant little or no replacement of equipment, and maintenance tended to be more corrective than preventive, occurring only after a breakdown had taken place. Some assets such as drilling barges, electric plants, refineries, tankers and terminals, and telecommunications showed clear indications of obsolescence. Modernization of equipment represented one of the major tasks facing the nationalized oil industry. Of the 14,000 closed wells, perhaps as many as 8,000 were capable of producing. The reactivation of these wells would be both expensive and time-consuming and would require millions of hours of analysis and repair work. The repair work would have to include flow lines and flow stations both on land and underwater in Lake Maracaibo.

In 1975 the industry owned eight drilling and ten work barges operating in Lake Maracaibo, but it was evident that more and more modern equipment would be needed in the short to medium term, especially for the drilling of deep wells in the southern portions of the lake.

Steam-injection plants were particularly important since they were associated with secondary recovery of heavy oils in the eastern side of the lake. Already very active in this area in 1975 was Maraven, with no less than six plants installed in the Tia Juana and Bachaquero fields east of the lake. Lagoven and Meneven had small units in eastern Venezuela which were temporarily closed in. The physical condition of these plants was reasonable. Many more would be needed in the medium term, given the increasing heavy-oil component of Venezuelan proven reserves.

Electrical plants were mostly very old; some had been in operation for more than three decades. The nationalized oil industry could not dispense with them, however, because the government-owned electrical network was extremely unreliable. The petrochemical plants of El Tablazo, at the north end of Lake Maracaibo, had been severely damaged by the frequent interruptions of electrical power supplied by the largest state-owned electrical company.

Of the twelve refineries existing in Venezuela at the time of nationalization, seven were quite small and obsolete, accounting for only about 15 percent of the total refining capacity. The large refineries of Amuay, Cardón, Puerto La Cruz, and El Palito were in good to excellent operating condition, but their yield had to be drastically altered to allow for the production of more medium and light distillates and of less residual fuels. The terminals associated with these refineries were mostly shallow water. Only the Puerto La Cruz terminal of Meneven, in Eastern Venezuela, could handle tankers of 150,000 tons. Seven of the thirteen existing terminals were located in Lake Maracaibo, and tankers went out of the lake through a navigation channel which was under constant dredging. This was the only way to permit the traffic of tankers up to 100,000 tons and was, and still is, an expensive operation not done by the oil industry but by an inefficient government-owned enterprise. Venezuelan terminals were at a disadvantage in relation to the deep-water terminals of Curaçao, Aruba, and Bonaire, which were built for supertankers. As a result the refineries in these islands could import Nigerian crude at a competitive price with Venezuelan similar light oils.

The tanker fleet was old and barely adequate. Of the twelve existing tankers, five had been in operation for more than 19 years and the rest for more than 13 years. Experience had shown that tankers in generally short runs, requiring frequent loading and unloading, were subject to much greater wear and tear than those engaged in transatlantic transport. As a result 18 years was probably a reasonable maximum for efficient use. On the basis of this criterion, it was obvious that most of the Venezuelan tanker fleet was in imminent need of replacement.

Another area of great concern to the newly nationalized oil industry was that of social facilities: office buildings and housing for personnel in the operational districts. Traditionally both the senior staff and the workers in those areas had been provided with low-rent housing by the companies. In a paternalistic fashion, facilities included low-cost electricity, running water, garbage disposal, and complete maintenance. Food stores carried a line of fifty to sixty basic items for prices which corresponded to those of 20 years ago in the "real world." With the advent of nationalization, personnel were expected to increase significantly with increasing activity. This would mean that sooner or later more offices, houses, schools, hospitals, roads, and other social facilities would have to be provided.

The Labor Force

In 1975 the oil industry had about 23,100 direct employees. 12,000 were white collar and fewer than 500 of these were foreigners. Ten percent of the total work force were university graduates, and about 3 percent were technical assistants. Specialized workers such as welders and carpenters made up about 25 percent of the total work force, and the remainder were nonspecialized workers and clerical staff. Work force was half of that in 1957, the last year of intense exploration activity in the country, when the industry had had about 46,000 employees.

The educational level of the labor force had been steadily rising. In the 1960s only 39 percent of the workers had primary schooling. By 1975 about 50 percent had it. Of special significance was the high proportion—25 percent—of employees with less than 5 years of experience, a problem which would get more acute within the next years, reaching a high of 45 percent in 1980.

From 55 percent to 60 percent of the total labor force was engaged in production work, the clearly dominant area of activity for the last decades. A very high percentage of the production workers were 45 years of age or older.

In general, then, the labor force was either very young or rather old. The older ones excelled at the occupations they had held for a long time but were rather poorly prepared to face the major changes about to take place. Morale was generally high, though, and so was the degree of confidence shown by the middle and upper managers in their ability to run the industry efficiently.

Technology

Another main area of interest was technology. A preliminary diagnosis of this field had been prepared in 1975 by a group of oil-industry technical staff under the coordination of INVEPET. The study showed that the Venezuelan component in the technological field was almost negligible at the time of nationalization. Most support had traditionally come from the multinational head offices but in some areas such as exploration, technology was essentially provided by service companies. The sector in which dependence from multinationals was probably the greatest was refining. In this field technology was very dynamic and tended to be developed within the research and development divisons of multinationals such as Exxon, Shell, and Gulf. Fields of research such as upgrading of heavy oil were of particular interest to multinationals since these types of oils would be more and more in demand as lighter-oil reservoirs became depleted.

AGROPET, the association of oil-industry employees, had also produced an analysis of the technological area. In their document three main types of oil technology were distinguished: research and development, licensed technology, and support technology. The first type they described as oriented toward the improvement of existing techniques, forming the basis for technological innovations. In Venezuela there was practically no oil research and development work to speak of. Only twelve full-time researchers in oil-related matters were active. Licensed technology was that already proven and in commercial use and was totally derived from foreign sources. Support technology was that which existed within each organization as the result of empirical trial and error, "the way things were done." The mature Venezuelan oil industry was particularly rich in accumulated experience which served to solve many day-to-day operational problems, to debottleneck operations, and to teach younger employees the tools of the trade. This type of informal technology was highly dependent on a constant inflow of new personnel who could bring their own knowledge into the common pot, to transfer new technology to the organization, however. Nationalization could shut off this inflow, with the inevitable long-term result of reduced efficiency and technical obsolescence. This was the main argument utilized by AGROPET to recommend that some type of technical agreements be negotiated with the main multinationals, those with enough technological capacity to be of real help to the nationalized oil industry. Oil-industry employees correctly sensed that agreements of this type would be needed, at least in the short term, until Venezuela developed other alternatives of technological support, including of course the creation of their own technological institute.

The Relation between the Ministry and the Nationalized Oil Industry

To understand fully the nature of the relationship between the bureaucracy of the Ministry of Mines and Hydrocarbons and the employees of the nationalized oil companies, it is indispensible to keep in mind the events of the early 1970s. During the period 1970–1974 ministry officials came to have an important voice not only in the traditional areas of technical auditing but in the less traditional areas of management. We have seen, in the previous chapter, how the combination of tax measures, the reversion and natural gas laws, and decree 832 gave government in general and the sectoral ministry in particular a very high level of control over the industry. Decree 832, issued in November 1971, was intended to clarify and regulate the provisions of the law of reversion. It required companies to submit in October of each year a copy of their operational programs for the approval of the ministry prior to their execution. These programs included exploration, production, refining, marketing, and capital budget items and could be studied and, if so desired, modified by ministry officials. For all practical purposes ministry bureaucrats had become the managers of the oil industry or at least its comanagers. In 1975 the ministry (today the Ministry of Energy and Mines) had several organisms through which it exercised this high level of control:

1. *The Coordinating Commission for the Marketing of Hydrocarbons* fixed the tax reference values for all crude oils and petroleum products. These were the values or prices on the basis of which companies had to pay income tax even if actual realization prices (the market price) were inferior. This tool assured government of a desired level of income independent of the fluctuations of the world markets. Price-setting was a totally unilateral decision on the part of the government, although there were frequent meetings between ministry officers and company representatives in which the latter would present their analysis of the international market and their price recommendations.

2. *The Joint Finance-Mines Commission,* with members from both the ministries of Finance and Mines and Hydrocarbons, regularly analyzed the financial results of the companies and the influence that income-tax payments had on those results. This commission was instrumental in the issuing of a decree which came into effect in January 1972 and which made mandatory for each company to export a certain basic volume of hydrocarbons during the year. If exports increased or decreased above or below these mandatory volume levels, the tax reference value or price would be subject to a variable surcharge proportionate to the extent of the volume variation. The psychological and material effects of these volume penalties were substantially negative as they forced the companies exporting below stipulated volume to lower prices in order to increase sales, further weakening the markets for Venezuelan crudes and products.

3. *The Hydrocarbons Division* controlled all the operational phases of the industry and, through the decree 832, was empowered to approve exploration programs, production levels, refining activities, and the amount of investment programmed for those sectors. The traditional tasks of this divi-

sion had always been of a highly technical nature, as it was overseer of the stipulations of the hydrocarbons law. After decree 832 was issued, these tasks became subordinate to those of a more managerial nature. For the first time in many years of rather silent and obscure service, the technical staff of this division tasted real power. They seemed to like it.

4. *The Reversion Division* was created in December of 1973 to coordinate the studies and the tasks which would be required to apply the provisions of the reversion law, including supervision of oil-company assets subject to reversion and control over the buying or selling of such equipment.

5. *The Local Market Programming Committee* regulated the supply and distribution of refined products in the Venezuelan domestic market.

These organisms did not usually work in a coordinated fashion. Many times orders issued by the Hydrocarbons Division would collide with orders from the Reversion Division. In these rather frequent cases oil-company management would invariably end up in a no-win position, since they could not satisfy all wishes. Rivalry between the different sectors within the ministry was strong, fueled by personal ambitions, mutual distrust, and even political motives. In 1975 the heads of the Hydrocarbon and Reversion Divisions were A. Reyes and H. Calderón Berti. These two engineers were both ambitious and relatively young. They belonged to different political parties: Reyes was a member of Acción Democrática and Calderón Berti a member of COPEI. Both wanted to become ministers, but the fulfillment of their ambitions depended on the degree of power they could now acquire. In the years 1973–1975 the ministry had all the legal tools for almost complete oil-industry control.

Both Reyes and Calderón Berti knew this and each started a strong drive to increase his personal power. At first Reyes had the upper hand as his Hydrocarbons Division was old, well established, and carried the prestige of many previous years of dedicated technical service. This division had representatives in all major operating districts of the country, men who had almost instant knowledge of the activities of the industry. The archives of the division dated back to the early years of oil activity in Venezuela. In contrast the Reversion Division had just been created; it had very little staff, a short budget and, in many ways, it had to create its own responsibilities as there was a significant amount of overlap between its terms of reference and those of the other departments in the ministry.

The difference was in the personalities of the two department heads. Reyes was more a typical ministry bureaucrat used to dealing with highly technical issues from a position of authority. He was not accustomed to participatory decision-making, but to giving orders. As a result he had not been able to develop the capacity for persuasion characteristic of good managers in large corporations. He lacked polish and was more feared than liked by oil employees. Calderón Berti was affable, highly articulate, and did not project an image of raw power, but one of genuine desire to cooperate. As a result he rapidly gained the acceptance of oil-industry personnel, and that seemed to give him the edge over Reyes.

Although styles were highly different, the underlying motivations were not. The struggle between the two department heads did little to improve the chronically low morale of ministry staff and greatly affected the normal work of the oil industry, which had to dedicate thousands of hours to provide both departments with essentially the same type of information in different formats. Not infrequently the information requested by one department would already exist in the files of the rival department but it would not be made available, in an effort to gain an advantage over the other.

It will be evident that the advent of nationalization and the creation of Petróleos de Venezuela, the holding company in charge of coordinating the activities of the operating companies, would introduce a new element in the relationship between the Ministry of Mines and Hydrocarbons and the nationalized industry. There was no longer a clear-cut situation in which government offices (the ministry) dictated to foreign companies (the concessionaires) what to do. Now there was a fully Venezuelan managerial body in charge of contributing the leadership. If this had been understood and accepted by ministry bureaucracy there would have been no conflict, because the ministry would have reverted to their original tasks of technical supervision. But since the desire for power was still a fundamental ingredient of ministry's bureaucracy, there was conflict, for managerial authority could not easily be shared by the two groups.

Ministry staff argued from the very first day of nationalization that decree 832 was still in force and that, therefore, all nationalized oil-industry activities had to have their previous approval. They argued that nothing had really changed in the relationship, when in fact significant philosophical and ownership changes had taken place.

Petróleos de Venezuela argued that the holding company had been created "to coordinate, supervise and control the activities of the nationalized oil industry" in accordance with the basic policies received from the National Executive through the assembly of shareholders. The Ministry of Mines and Hydrocarbons still possessed all faculties given to it by the hydrocarbons law, essentially those of supervision and technical auditing of the operational activities of the industry. Petróleos de Venezuela on the other hand had the responsibility of managing industry so as to fulfill the basic objectives its shareholders had asked it to pursue. Decree 832 therefore unnecessarily duplicated the effort involved in the analysis of budgets and operational programs. A managerial concern and participation of ministry staff in the activities of the industry was inefficient and would inevitably lead to confusion and friction. In the opinion of Andrés Aguilar, the legal advisor of Petróleos de Venezuela, decree 832 had "automatically been rendered invalid" by the law nationalizing the oil industry.

But ministry officials had a different view. For many of them the continued presence at the helm of the nationalized oil companies of the managerial group that had worked under the multinationals seemed to be enough reason for distrust.

A high-level officer at the ministry once confessed this feeling to me. He said, "We don't know everything that is going on in the industry and I have the suspicion that it is in those areas we don't know much about where you are deceiving us." This deeply ingrained distrust was probably the product of the many years in which the government sector had in fact been left very much in the dark by more experienced oil-industry staff.

A meeting held at headquarters of Petróleos de Venezuela in April 1976, in which all the top executives of the ministry and of PDVSA were present, illustrates the frictions existing between the two groups.[2] This meeting essentially dealt with two items of vital interest at the time: the exploration program and the financial situation of the industry. Concerning the exploration program, Arévalo Reyes stated that operating companies still had much ground to explore in their traditional areas. He said that the evaluation of new areas would be done by the ministry, "since the operating companies had not been able to do a proper job." PDVSA replied that the creation of an exploration group in the ministry to evaluate and manage the new areas would be wasteful of the limited human resources Venezuela had in this field and would collide with the concept of the ministry as a technical auditor and of PDVSA as the managerial and technical coordinator. In this we were supported by Calderón Berti, who said very emphatically that the job of the ministry should not extend into the field of operating and managing the industry.

Discussing the finances of the industry Pablo Reimpell, from PDVSA, explained that because of the wide gap between the tax reference prices and market prices, a situation existing since the second half of 1975, PDVSA and the operating companies would show a corporate loss of about Bs 400 million (about $90 million). He suggested that a reduction of 10 percent in the tax reference prices would eliminate this loss and give industry a corporate profit of about Bs 600 million. General Alfonzo added that the industry could not be expected to pay in taxes 110 percent of its income, since this would lead to its decapitalization, bankruptcy, or politicization. Dr. Altuve, from the ministry, said that "it was not required for the industry to make a profit, but to obtain from the government the funds they needed." The minister added that he did not favor a reduction in the tax reference prices, and that some other solution should be found.

It is evident that there were major differences of outlook between the two groups. For the industry, to show a corporate loss during its first year as a state-owned enterprise would be a major psychological setback since very few people would understand the reasons for this. Public opinion would tend to equate the results with mismanagement under the new system.[3] It is surprising that oil-industry technocrats seemed to be, at this time, much more sensitive to the political implications of the situation than the more politically oriented ministry bureaucrats.

Notes

1. J. Méndez and J.R. Domínguez, "La Exploración para Hidrocarburos en Venezuela," Maracaibo, 1975, p. 77.

2. Attending the meeting for the ministry were Minister Valentín Hernández, the Vice-Minister. Hernan Anzola and Directors Arévalo Reyes, Humberto Calderón Berti, and Francisco Gutierrez, accompanied by Advisers Altuve, Méndez, and Lauder. From PDVSA attended General Alfonzo and most of the board members.

3. Tax reference prices, it will be recalled, were those values set by government to base income-tax payments from industry. If these prices were much higher than market prices, it was possible that operating companies would have to pay in taxes more than 100 percent of their income.

7

Some Major Early Issues, 1976–1978

From Fourteen to Four Operating Companies

The fourteen operating companies created at the time of nationalization were too different in size, technical know-how, and strength of labor force for all to be able to survive. From the beginning there was a strong intuition that the number of operating companies should be drastically reduced. President Carlos Andrés Pérez had already advanced a scenario in which only four companies would remain: Lagoven, Maraven, Meneven, and a fourth one to be created by fusing all the other companies together.

In 1976 Petróleos de Venezuela (PDVSA) formed a committee of the board made up of J.G. Arreaza, J.D. Casanova, L. Plaz Bruzual, J.R. Domínguez, and myself to study the ways in which a process of reorganization of the operating companies could be accomplished. From the early stages this committee had the technical cooperation of a management consulting firm (McKinsey and Company). The committee decided to conduct interviews with the top executives of every one of the operating companies so as to hear their opinions on how to simplify the operation of oil companies. This process took several weeks but was fruitful, since it gave the committee several ideas on how to proceed. It also allowed the executives of the different operating companies to express their own views on an issue not at all easy to resolve to everyone's satisfaction.

The issue was in fact very difficult. Although most of the top oil executives were men of intense loyalty to the industry, they also had a deep sentimental attachment to their own organizations and naturally wanted them to survive. Some soon realized, however, that this survival would not be in the best interests of the industry. It was to their great credit that almost without exception they correctly perceived the constraints and cooperated with unfailing dedication in a task which, for many of them, meant a demotion and the disappearance of the organizations they had helped to create and to which they had become so attached.

In July 1976 the presidents of the five largest operating companies—Lagoven, Maraven, Meneven, Llanoven, and CVP—held a series of informal meetings on this subject and produced a document which carried much weight since it contained the consensus of the chief executives of those companies and advanced very concrete ideas on how to attain the objective of rationalizing the administrative structure of the industry. This document advanced the following conclusions:

1. It is urgent to reduce the number of operating companies reporting directly to Petróleos de Venezuela.
2. This objective can be achieved immediately through the concept of larger companies serving as advisory agents to smaller companies, which, through this mechanism, would not lose their corporate identities. The advisory role of the larger companies would not imply that they could intervene in the day-to-day administration of the smaller companies.
3. To move systematically toward the final objective of having only four operating companies, both operational compatibility and managerial capacity should be taken into account as primary criteria to fuse companies together.

The document went on to propose the following scheme for the first stage of "coordination," as they called it:

Lagoven	Maraven	Llanoven	CVP	Meneven
Amoven	Palmaven	Bariven	Boscanven	Guariven
Roqueven		Vistaven		Taloven
				Deltaven

The document also contained a proposal for a second stage, "integration," in which Lagoven, Maraven, and Meneven would remain essentially intact, and CVP and Llanoven would merge into one company. The presidents stressed that the maximum number of operating companies should be four.

This document was received with mixed feelings by some members of the board of PDVSA and by the presidents of the smaller companies. Although they conceded that such an initiative had been taken with the purpose of coming up with a sensible simplification scheme, they could not help feeling that the bigger fish were simply getting together to decide the fate of the smaller fish. Some of the members of the board of PDVSA perceived the document as an act of power, while the presidents of the smaller companies essentially felt left out of the decision-making process.

Some members of PDVSA's reorganization committee, aware that this could develop into an area of major friction among oil industry executives, decided to conduct more frequent meetings with the presidents of the smaller companies. At their request General Alfonzo had a first meeting with the nine presidents of those companies. This meeting did not work out well, however, since General Alfonzo bluntly told these executives that their companies were too small to be of consequence—a bitter truth for anyone to hear.

The analysis on reorganization prepared by the committee was based on sixteen working papers on the subject presented to PDVSA by the presidents of the operating companies and by two directors of PDVSA; on the minutes of meetings held with all of the presidents of the operating companies; on the ideas presented to the committee by dozens of oil-industry executives; and on conversations held within the committee.

Basically the document suggested that existing operating companies could be divided into five groups:

Companies with substantially large organizations—Lagoven, Maraven, Meneven

Medium-size companies, which would need to be significantly strengthened—Llanoven, Deltaven, Palmaven

Small companies, which should be integrated into larger organizations in the short term—Boscanven, Roqueven, Bariven, Amoven

Marginal companies, which should be integrated into a larger company at once—Vistaven, Taloven, Guariven

CVP, the original state oil company.

When basic industry parameters were considered, it became evident that the first group of companies had 85 percent of the proven oil reserves, 80 percent of the production capacity, 84 percent of the refining capacity, and 72 percent of the labor force. Only five of the companies: Lagoven, Maraven, Meneven, Llanoven, and CVP were substantially integrated.

It was clear therefore that the companies with the most capacity to survive and incorporate others into their organizations were Lagoven, Maraven, and Meneven, and that Llanoven and CVP were borderline cases. Based on an evaluation of the operational, managerial, and labor aspects involved in the analysis, the document recommended the following coordination stage:

Coordinating Companies	*Lagoven*	*Maraven*	*Meneven*	*Llanoven*	*CVP*
Coordinated	Amoven	Roqueven	Vistaven	Bariven	None
		Palmaven	Taloven		
		Boscanven	Guariven		
			Deltaven		

This document was distributed to all the members of the committee. During the two presentations made to all committee members, there were no objections. As a consequence of this tacit approval, a presentation was made to the executive committee of PDVSA in September 1976. At this presentation one of the members of the committee insisted on his original ideas, which had been in the minority at the working committee level. Since he was a senior director, however, his strong disagreement at the executive committee meeting forced a revision of the issue. As a result there were some changes made to the original recommendations which had suggested that the ideal number of operating companies should have been only three. Lagoven, Maraven, and Meneven, and that CVP should have been converted into a company to take care only of the local hydrocarbons market. This recommendation proved to be politically unacceptable since CVP had been, in the eyes of large political sectors, a symbol of true nationalism.

CVP was a rather weak company, which had been substantially politicized during the last five years immediately before nationalization. After the departure of R. Sáder Pérez, the presidency of CVP had gone to mediocre leaders who had been mostly content with survival. In 1976 CVP had only 4 percent of total industry production, but 13 percent of the work force due to the hypertrophy of its administrative services and to its responsibilities in the domestic market.

The modified reorganization was approved by the board of PDVSA in September 1976 and sent to the Minister of Mines and Hydrocarbons Valentín Hernández, who also approved it in November of the same year.

The process started in January 1977 and, by the end of the year, the smaller companies were already being absorbed into the larger ones. In 1978 the fourth operating company was created with the name of Corpoven (basically a fusion of CVP and Llanoven) and the process was essentially completed. As General Alfonzo Ravard said in his speech at year's end in 1978:

> The rationalization of the petroleum industry is a gradual process that has been advancing without the slightest interruption in our oil production. . . . During these three years (1976–1978) the oil industry . . . has been able to acquire more global vision and . . . uniform systems and norms.[1]

What was really noteworthy about the consolidation of the fourteen original operating companies into four large organizations was that it took place without a break in operations.[2]

A result of the rationalization process was the increasing uniformity of administrative systems and procedures. Up to that period all companies had had their own set of personnel, finance, and general administrative systems. Now, with the great volume of personnel moving across organizational boundaries, it became necessary to establish common norms and procedures. Work to that effect started in the field of job description and evaluations since people being transferred from one organization to another had to fit into the new system without suffering a deterioration of the working conditions they had been enjoying in their previous jobs. The HAY-MSL job evaluation system was globally applied within the industry. Procedures were created to guide intercompany transfers in order to minimize competition for the same human resources. Great efforts were directed toward the creation of more unified corporate climates without destroying the individual work styles of the organizations. Companies eventually adopted certain procedures from others and the industry in general started moving closer together after many years of intense rivalry and secrecy. For the first time in their long history, the organizations undertook a widespread exchange of technical and administrative information. This greatly contributed to the improvement of managerial quality within the industry. During the first half of 1976, the holding company became essentially dedicated to the structuring of the different coordinations and to deal with urgent policy issues and projects such as jurisdiction.

Jurisdiction Over the Orinoco Heavy-Oil Project

There were deep differences of opinion between the Ministry of Mines and Hydrocarbons and the oil industry regarding the Orinoco heavy-oil deposits and the ways to develop them. The ministry's technical staff wanted to retain operational control of this project, while the industry felt it had all the resources to take over operations and conduct them in a more efficient manner.

In February 1976 the Caracas newspapers reported that Oil Minister Valentín Hernández had mentioned the possibility of negotiating pilot projects in the Orinoco area with the governments of Canada and Rumania, as well as with some American multinationals.[3] This report was brought to the attention of the board of PDVSA by Director Julio Sosa Rodriguez, a very outspoken critic of foreign participation in the development of the Orinoco heavy-oil deposits. Sosa Rodriguez argued that such negotiations would be inconvenient since foreign concerns would certainly seek participation in the future production of the heavy-oil reservoirs. Director Carlos Guillermo Rangel agreed with Sosa Rodriguez and added that it was important to clarify the position of Petróleos de Venezuela in this issue. Andrés Aguilar, PDVSA's legal advisor, suggested that efforts be made to define the jurisdiction of the ministry in the management and operational control of the petroleum industry. The board felt the project should be transferred to PDVSA for management. During several months of 1976 tension grew

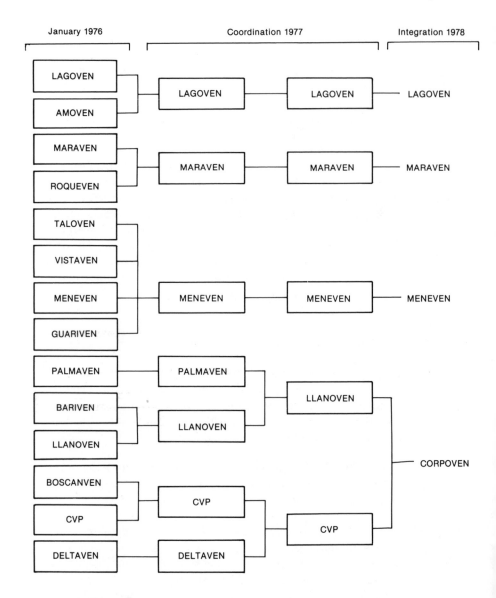

Source: Petróleos de Venezuela, *Informe Anual,* 1979.
Figure 7–1. The Process of Industry Consolidation

between the ministry and PDVSA over this issue. In April 1976 Julio Sosa Rodriguez resigned from the board in a move perhaps partially induced by his concerns about the Orinoco project and his impatience over the lack of a clear policy on this issue.

Oil Minister Valentín Hernández favored handing the Orinoco project over to PDVSA, but he was hard pressed not to do so by his militant staff at the ministry and by some political sectors which insisted that giving the management of the Orinoco project to PDVSA was tantamount to handing it over to the multinationals. The work of the Permanent Commission of the National Energy Council is a good illustration of this tendency. This group had very vocal members, who demanded the Orinoco project be handled only by "patriotic" Venezuelans such as those working at the ministry. PDVSA felt therefore that a showdown was inevitable and went to President Pérez requesting a decision on this matter.

The memorandum presented to President Pérez by PDVSA stressed the following arguments. The Orinoco heavy-oil reservoirs differed little from those already being operated by the industry in the nearby Morichal, Jobo, and Pilon oil fields. PDVSA was the organism created by the state to program, plan, and conduct activities in the oil sector. The Orinoco project was simply one more project within this sector and should not be managed by another organism in an entirely different fashion. The organization created within the ministry to manage the Orinoco project was small and lacked the financial, technical, and human resources required to do the job properly. The research or experimental portions of the project would be better handled by INTEVEP, the research and development Institute of the industry, than by outside contractors, the way the ministry proposed to do it. Integral planning of petroleum activities could not be done by the industry without incorporating the Orinoco project.[4]

In time, the power struggle was decided in favor of PDVSA. The Orinoco project organization of the ministry was dismantled and all information transferred to PDVSA. According to INTEVEP and PDVSA staff who examined this information, the 5 years of evaluation work done at the ministry had proved to be an almost complete waste of time.

The Bioprotein Project

In 1974 the Ministry of Mines and Hydrocarbons contacted British Petroleum, indicating an interest in constructing a plant to manufacture bioprotein. Proteins for animal feed were to be derived from hydrocarbons by means of a technique patented by British Petroleum in the 1960s. Bio-Proteinas de Venezuela was incorporated in June 1975 with capital contributed by the government (60 percent); British Petroleum (20 percent) and

AFACA, the association of Venezuelan manufacturers of animal feed (20 percent). Original capital was Bs 2 million; the government share was contributed by CVP, the state oil company, according to instructions received by CVP from ministry officials, especially the director of the Hydrocarbons Division. In June 1976, with the oil industry already under the administration of PDVSA, CVP approved an increase in the capital to Bs 15 million and only later informed PDVSA. Grave doubts already existed in PDVSA about how the project was being managed and about the adequacy of government representation on the board of Bio-Proteinas. Although two of the members of the PDVSA's board argued strongly against approving this increase, the majority decided to go ahead with the project.

The need to increase the capital was never explained satisfactorily to PDVSA, but this was not really surprising since for the last two years there had been significant expenses connected with frequent trips by government officers to visit bioprotein plants in Europe, as well as with a number of feasibility studies. In November 1974, for example, a group of seven persons traveled to Scotland, France, Italy, Switzerland, and Holland. Feasibility studies were commissioned to Stone and Webster; Parra, Ramos, and Parra; Davy Powergas; Planesa; Battelle, and others. In February 1976, another group traveled to Italy. Bidding was opened for the construction of a paraffin plant and a protein plant. The shareholders of Bio-Proteinas agreed, in March 1976, to commission these plants to Technipetrol and Stone and Webster, respectively, although no licensing contract had yet been signed with British Petroleum. Expenditures had already exceeded the available capital. In travels and salaries alone almost Bs 1 million had been spent in 1975. Office space for the company was bought in Caracas for another almost Bs 1 million. Yet another million went for several preliminary studies. When the capital increase was requested, the obligations of the company already amounted to Bs 12.5 million.

The approval of PDVSA was given, but with the conditions that it would not imply blank approval for further capital increases, that the work assigned to Technipetrol and Stone and Webster stop at a certain cost limit and that it be discontinued beyond this stage, that the board of Bio-Proteinas refrain from handing out further contracts. Despite these stipulations, the Bio-Proteinas management group eventually increased indebtedness to about Bs 50 million before the project was finally mothballed by PDVSA, a classic example of bureaucratic confusion which ended with CVP, the Ministry of Mines and Hydrocarbons, the Venezuelan Investment Fund, and PDVSA bitterly arguing and blaming one another.

The Petrochemical Institute

Petrochemicals in Venezuela date back to 1954 when the Pérez Jiménez military dictatorship established a petrochemical master plan, based on the reasonable assumptions that natural gas was plentiful and that oil should be upgraded to be converted into fertilizers, explosives, and plastics in order to become truly valuable. The site chosen for the "Petrochemical Institute" was Moron, near the sea, on a property owned by one of the close friends of the dictator. The original complex consisted of a small, 2,300-barrels-per-day refinery, a fertilizer complex, plants to produce caustic soda and cloro-soda, and a 15-megawatt electric plant. From the start the plants did not add up to a real petrochemical complex, and their technology was semi-obsolete. When the Pérez Jiménez government was overthrown in 1958, the new administration of the Petrochemical Institute made an analysis of existing operations and concluded that the institute was in serious trouble: the personnel were highly demoralized. Most of the plants were producing at only minimal capacity and with imported raw materials.[5] The new administration did no better, however. From 1959 to 1977 the petrochemical industry suffered a succession of failures brought about by managerial mediocrity, labor-union irresponsibility, and political interference. The predominant approach to planning was essentially political, considerations of this type often deciding personnel policies. A former head of the institute, E. Acosta Hermoso, admitted to these practices when he wrote that

> We wanted the investments rapidly approved because any delay would mean a deferment of construction and the partial failure of our government program. We had 5 years to prove we were capable, so as to be able to win the next elections.

This statement looked like a typical example of the political cycle prevailing over long-term, careful planning. He added,

> Once, in the face of heavy criticism from outside sources, I asked the Ministry for authority to remove 600 workers and got a rotund negative answer. I complied, knowing that this would save (the institute) from political and social conflict.[6]

In other words the manager decided not to do what he must because of the short-term political impact of his decision. The results of this management, combined with featherbedding and corruption, was easy to predict. By 1977 the accumulated losses of the petrochemical complex were of some Bs 4 billion and personnel were highly demotivated. As a measure of last resort the government decided to turn the institute into an affiliate of PDVSA. This decision was very risky and was accepted by PDVSA only with great misgivings as there was a great danger of contamination of the healthy oil industry by the extremely ill petrochemical industry. PDVSA insisted on taking over the petrochemical industry only if the government agreed to cancel its debts. The government agreed to this and, as a result, the Petrochemical Institute became an affiliate of PDVSA in March 1978 under the name of Pequiven. After the takeover, the petrochemical industry reduced losses from some Bs

500 million in 1977 to only Bs 62 million in 1981. But PDVSA has been obliged to make important contributions of capital to Pequiven, a total of about Bs 1.5 billion in the years 1980 and 1981.

Exploration Planning

The generation of new oil reserves was one of the major tasks facing the newly nationalized oil industry. The lack of sufficient experienced personnel was, however, a major constraint for the conduction of a large-scale exploration program. When the exploration coordination in PDVSA was formed and a coordinator had to be named, the job went to Alex Lorenz, an explorationist of great experience who had been, up to that moment, exploration manager of Lagoven. He was a mature and dedicated geologist but had the belief that most of the exploration studies should be conducted at the level of the holding company. This belief led him to present to the executive committee of PDVSA an organization proposal which would have resulted in the hiring of about sixty persons in the near term.

This proposal worried some of the members of the board as well the top executives of the operating companies, who felt that, if approved, it would cause a massive transfer of exploration personnel away from the operating companies, weakening those organizations and leading to an undesirable concentration of manpower in PDVSA. The presidents of the operating companies met informally with some of PDVSA's directors and together produced a document calling for the creation of exploration task forces to conduct the evaluation of those areas which could be subject to immediate exploration, both inland and offshore.

The idea of creating task forces to do the job was far more sensible than that of creating a large exploration bureaucracy in PDVSA. The board accepted this recommendation, not without great infighting, and the task forces were structured. The composition of these task forces depended on the knowledge each operating company had about the areas to be evaluated: Maraven received the responsibility of coordinating the evaluation of the offshore areas of western Venezuela, CVP those of central Venezuela, and Lagoven those of eastern Venezuela.

The exercise proved to be very profitable in every way. The work of the different groups helped to outline the major areas of exploration interest in the country, especially offshore. It led to the formulation of a systematic exploration campaign and helped to create an atmosphere of cooperation among the members of the different operating companies. Traditionally exploration work had been conducted with utmost secrecy. As a consequence there had been minimal exchange of information among organizations. Staff belonging to different companies rarely spoke to one another about their own work. Now the scientific files were opened for everyone to see. Geological and geophysical data were compared and views freely exchanged. The recommendations which finally emerged were the product of the study of the best information owned by all organizations. On the basis of the recommendations made by the various task forces, three offshore drilling rigs were contracted on a 2-year, renewable basis, and offshore exploration began in 1978.

The Evolution of the Work Style at PDVSA

With the resignation of Julio Sosa Rodriguez, the board lost the only member who could occasionally argue successfully with General Alfonzo. The meetings became set in a routine in which Alfonzo would expand on his views without much participation from the audience.

A particularly negative characteristic of many of PDVSA's top executives was their lack of punctuality. PDVSA's offices were on the eighth floor of the Lagoven building and the parking area closest to the building had been reserved for its directors and other high officers. Lagoven employees, coming to work in the morning, would see this area mostly empty. This made a very negative impression since, in the oil industry, most employees arrived to work on time. On the other hand ministry and government-agency employees were not so careful about being punctual. Lagoven employees probably began to consider the empty PDVSA parking spaces as a sign that the philosophy of management in the holding company was going to be closer to that of a government agency. They began to worry.

There were also positive signs. General Alfonzo successfully kept politics out of the industry. He handled most situations of this type with the great skill acquired in his long years as a top manager of public enterprises. He seemed to give in to the small things, like having to give a $200,000 contribution to a political government exhibition, labeled by Caracas humorists as "King Kong's cage" because of its peculiar layout, or hiring an inept former ambassador to Holland as chief of protocol only because he had been sent to PDVSA by some top government officials. In general however, General Alfonzo was highly successful in keeping the oil industry free from large-scale political interference. In this he was enthusiastically assisted by most of the top executives of the industry, who dreaded the day when they would start getting orders from the government bureaucracy.

The strategy of Alfonzo Ravard was clearly a good one since it prevented for several years the politicization of the Venezuelan oil industry. What is not so clear and now might be purely academic to consider, is whether a more rigid attitude on his part from the very beginning could have prevented politicization for an even longer time. Hiring an inept man for a subordinate position only because he is a friend of government bureaucrats might seem a peccadillo when compared to worse things that he was able to prevent, but it indicated to members of the political world that here was a man who could be "pragmatic."

At the end of 1977 Petróleos de Venezuela had acquired many small bad habits and committed many peccadilloes. The company was starting to deviate significantly from the oilman's vision of what a business concern should be. Some points of deviation were as follows:

Some of the members of the board dedicated themselves to the structuring of small enclaves of personal power instead of contributing to solving the overall problems of the organization. This occasionally led to the recruitment of substandard personnel in an organization that did not need quantity but very high quality. As a result the holding company was not perceived by the industry as its leader.

The holding company lacked an efficient information system; often the right hand did not know what the left hand was doing. Organization was loose in many ways. Meetings were incessantly held, but they rarely started on time and rarely followed the agenda. Every coordinator seemed to have his own ideas on how to organize his group. Since there was no corporate organizational philosophy, the operating companies often received contradictory signals. Several high-ranking officers, such as the legal advisor, were not working for the company on a full-time basis.

Many important decisions on recruiting new employees, on organizational matters, oil prices, and investments were made without sufficient analysis, in an intuitive manner. Moreover, important matters such as the reorganization of the industry were no longer subject to open discussion, decisions being made by a small group of people.

Some final, smaller points of contention: The holding company had created some agencies abroad which were expensive to maintain and which did not seem to fulfill worthwhile missions. Coordinating departments were proliferating and some, such as Computing and Urban Planning, did not appear to be justified. Advertising expenditure by PDVSA was excessive and all business given to an agency which had shown little talent.

The Interface with the Operating Companies

Operating companies were having a difficult time. They were being pressed from all sides: the need to carry on efficient operations, the need to absorb the smaller organizations efficiently and untraumaticly; the adjustment to the new boss, Petróleos de Venezuela. The Ministry of Mines and Hydrocarbons kept making demands on them as if the concessionaires were still in the country, as if nothing had changed.

The overriding concern of the top executives in the operating companies was clearly to keep the industry strong. To do that, they felt they should have full authority to make major decisions. They knew that full authority was possible only if PDVSA remained a rather remote holding company, letting them do their job without interference. The way to ensure that this would happen, they felt, was to staff PDVSA with employees who would tend to call the operators to ask for support rather than to give them orders. They also tried at times to keep some of these employees feeling that their main loyalty was to the corporation they had come from, more so than to their holding company. This was of course the wrong attitude and fortunately did not become standard practice.

Operating-company executives felt that the burdens of dealing with the Ministry of Mines and Hydrocarbons should be taken off their shoulders by PDVSA. In many respects, this was done, but freedom was far from complete. The interface between ministry officials and operating company employees was great in the oil fields, and that was where most problems occurred. Oil production could be ordered shut by ministry inspectors who wanted to exercise their authority, just as in the old concessionary times. Houses and company cars could be requested by the local hydrocarbons chief over and above their agreed allocation. The overall quality of ministry technical staff had been steadily deteriorating over the years, and much of the field staff was composed of those engineers who had not been recruited by the oil companies. Since they were modestly paid as compared to their counterparts in the companies, they tended to compensate by using their official status and to impose their will over the companies.

There was no doubt therefore that a new dimension had been added to the work of the operating companies. They now had a great technical challenge, an immense administrative challenge, and they also had a new, substantial political challenge. This new political ingredient had come about as a typical result of a nationalization. Before, the oil industry had been a rather isolated sector, going its own way without being an integral part of Venezuelan life. Now it was part of it, and everyone in the country suddenly felt that they had a voice on how it should be run.

Notes

1. Alfonzo Ravard, "Cinco Años de Normalidad Operativa, Petróleos de Venezuela, 1975–1980," Caracas, p. 190.

2. *Resumen* 390, April 1981.

3. *El Nacional,* February 7, 1976, front page.

4. This summary is based on a preliminary version of the PDVSA document.

5. E. Acosta Hermoso, *Petroquímica, Desastre o Realidad?* (Caracas, 1977), p. 67.

6. Ibid., p. 126; pp. 147–148.

8 Organizational Changes and Operational Expansion, 1976-1979

The Evolution of the Organizations

After years of producing and refining, the Venezuelan oil industry had become a place for operators, for doers. Good management had become identical with cost control. With nationalization there was an immediate need for long-range projects in all fields and the handling of these projects required a different breed of oil employee. It required planners as well as operators and the corporate view in addition to cost monitoring. It required, in fact, an attitude toward spending that industry management had largely lost after so many years of restricted budgets and as low as possible operational expenses. The task of the top executives at the operating companies was to change the mood of their organizations, to stimulate new ideas, and to let personnel know that once again they were free to generate projects and ideas even if these ideas departed from the orthodox.

The primary problem industry had to face in the field of human resources was where to find the people required, people with the expertise and attitude needed to deal with the new business environment and the new objectives of the oil industry. The most urgent consideration was to survey internal resources, since organizations tend to have significant resources which remain untapped in times of contraction and which can be efficiently utilized in times of expansion. The most important quality for these human resources to have was an outstanding capacity to adapt to new situations, a liking for change, as opposed to the more conventional employee or manager more intent on preserving the status quo. These resources had to be identified and placed in positions where they could freely develop their abilities and serve as the agents of change.

The expanding industry became fertile ground for the personal growth of those employees who were attuned to the changing atmosphere. This does not mean that the more tradition-oriented employees had no role to play in the postnationalization organizations. They acted as a moderating force, providing checks and balances to the drive for change since there were many excellent reasons to preserve many of the existing ways.

The second source of personnel was the outside world, new graduates as well as managers and technical staff from other industries or from public service. The industry suddenly realized that it did not have an efficient recruiting system. Since for many years recruiting had been at very low levels, it had been left to the functional managers themselves, each manager making his own contacts, interviewing and evaluating. Now, with the need for a much larger recruiting effort the companies realized that new, professional departments had to be created. This posed a very interesting problem, since many of the first recruits would have to be precisely those needed to staff the recruiting departments. Recruiting, induction, and training groups had been largely dismantled in favor of industrial and labor relations, remuneration, and employee benefits. Personnel managers were almost exclusively labor relations experts who knew or cared little about recruiting, induction, career planning, organizational development, and the subtler problems of organizational behavior which would be posed by the significant growth of their organizations. The problem faced by industry in the field of recruiting was also present in all other areas of activity. It was related to the ability of organizations to outgrow themselves, to break away from years of habit and routine into new ground where creative thinking and modern management techniques would be indispensable.

It took some top executives little time to realize that the new tasks required much more than a simple increase in recruiting. They also required a new, streamlined, lean organization with quick reflexes, as opposed to the rather slow-moving companies of old. However, at the top executive levels there were also problems. With the exception of Maraven and Lagoven, which had boards long in experience and managerial talent, the other companies were in bad shape. Meneven had been forced to promote to the board a group of young managers who did not have the necessary knowledge or influence to make of Meneven an equal member of the club. As a result Meneven was treated by the larger affiliates and by PDVSA as a younger sibling or child in constant need of support. The Meneven board accepted this situation and, in fact, seemed to encourage it by constantly leaning on other organizations, including the Ministry of Energy and Mines (the former Ministry of Mines and Hydrocarbons), to obtain reassurance for their decisions. Corpoven was in the process of being created, the result of a merger between CVP (the original state oil company) and Llanoven (formerly Mobil), with the addition of Deltaven (ex-Texaco) and Palmaven (ex-Sun). Although the average experience of top management in this company was significantly greater than in Meneven, they also had acute problems. CVP had been badly politicized, especially during the 6 years immediately prior to nationalization, and as a result the quality of managers and the morale of the organization were very low. Llanoven had more managerial talent, but it was coming into the merger as a subordinate entity, mostly due to political considerations. CVP had always been a political symbol just as national airlines or a soccer team can be political or status symbols. It had represented the original fortress of nationalism in a field dominated by foreign influences and by Venezuelan managers who had been working for foreigners for many years and who were therefore not to be entirely trusted. If CVP had to merge, went conventional political wisdom, it should keep a dominant position. The emergence of CVP as the dominant partner in the merger which led to the creation of Corpoven was clearly the result of a compromise between administrative common sense and traditional Venezuelan politics.

Organization Studies

The top oil executives who had recognized that expansion was not a simple matter of recruiting were essentially those at Maraven and Lagoven, especially the former. Maraven undertook a study of its organization made by an American consulting firm (Tower, Perry, Foster, and Crosby, a New York-based company) as early as 1977. In this analysis Maraven was perceived as a very good company with certain particularly weak areas:

> The members of the board were strongly functionally oriented, each director almost exclusively concerned with his immediate area of responsibility. As a result, they only had remote corporate concerns.

> The presidency (president and executive vice-president) was not working as a totally smooth team.

> First-line management lacked depth as a result of expanding activities and of loss of a great number of Maraven executives through transfer to PDVSA and other companies.[1]

> Managerial information systems were obsolete and needed urgent revision.

> Personnel procedures and norms had to be revamped and personnel staff greatly strengthened.

> All planning activities needed to come under the responsibility of a corporate director who would work intimately with the presidency.

As a result of this study, a rapid organization change took place in Maraven during the period 1977–1979. Personnel increases were carefully monitored and subject to continuous revision. This careful approach to recruiting made Maraven the company with the slowest rate of growth during those years. The role of the board of directors changed substantially, from an almost purely functional orientation to a group with strong corporate concerns. The transition was painful since several of the directors had for years been strong functional managers and either did not wish to become involved in other aspects of corporate life or did not have the adaptability to do so. The president and the vice-president started meeting regularly and acting more as an interchangeable team. Careful ranking exercises were conducted to select first-line managers who could occupy the new positions defined in the study. A managerial information system came under the responsibility of a corporate director, and some improvement (not nearly enough) was made in modernizing personnel norms and procedures.

The example set by Maraven in conducting an integral organizational study was not followed by the other operating companies until much later. The studies of Lagoven and Meneven were executed some years later, in 1980 and 1981, whereas the organizational study of Corpoven was only completed in 1982. This does not mean that these companies were totally unconcerned about this major issue. Lagoven, especially, had always been careful of its organization and, when expansion was undertaken, from 1977 onward, its attention was focused on getting the job done. However, Lagoven had not paid equal attention to other aspects of great corporate importance such as medium- and long-range planning; the integration of major project management at board level; the interface with the external sectors under the new, nationalized conditions; personnel planning; the link between finance and planning; the resolution of interdepartmental conflicts. As a result, Lagoven grew very rapidly, some big projects ended up costing considerably more than originally planned, and its handling of the Orinoco (Cerro Negro) project was subject to considerable criticism from the political and bureaucratic sectors which did not want to see the Orinoco heavy-oil deposits developed by PDVSA. Although Lagoven survived the expansion years in good style, it dealt with its challenges in a much less anticipatory manner than Maraven did. It was not until the end of 1981 and 1982 that the recommendations of the consulting firm (McKinsey and Company) were put into effect. Recommended were the creation of a board planning committee; strengthening of the planning staff; establishment of a manpower planning department, including a personnel-development planning group; organization of major projects group and establishment of a project-management system; and finally development of a more sophisticated finance function, "to include a more forward-looking, analytical role."[2]

Meneven also completed an organizational study in 1981. Many of its findings were already very well known both inside and outside the organization. Mene Grande, the predecessor of Meneven, had been—of all the large operating oil companies based in Venezuela prior to nationalization—the least autonomous. Owned jointly by Gulf, Exxon, and Shell, it had been little more than a production division with yearly targets rather than objectives and with a budget rather than with a plan. As a result its board was a group of functional managers without a corporate presence, much more so than in the case of Maraven. The production divisions in San Tomé and Lagunillas were too strong and acted essentially on their own without being guided by corporate objectives. The staff functions such as planning, finance, and human resources were very weak and needed urgent strengthening. Major projects were not coordinated by a project management division but by the individual functions. The credibility of the company vis-à-vis PDVSA and the other operating companies was extremely low. Managerial talent was in very short supply as there had never existed a systematic personnel development program.

Meneven was essentially dominated by the production function. Production managers and the production director therefore had an exaggerated influence in overall decision-making. There was widespread cronyism, with rewards and penalties often tied to personal loyalty rather than to company loyalty. For the last years prior to nationalization and, then, after nationalization some Meneven employees had developed strong personal ties with members of the Hydrocarbons Department of the Ministry of Energy and Mines. The production director at the time of nationalization had been a member of the ministry staff and was extremely sympathetic to the efforts made by the ministry bureaucracy to retain managerial control of the industry. In many instances, Meneven would look up to the ministry directives in preference to PDVSA's. This had contributed to the poor credibility that Meneven had within the industry. The mounting problems within this company led PDVSA to conduct a major overhaul of its board in 1980, naming three of its members from outside the organization, an unprecedented move.

Corpoven also had acute problems. The product of a merger of diverse organizations, this company spent most of the 1976–1979 period merely surviving its many operational and organizational crises. The CVP culture, akin to that of a government agency, came together with the Mobil, Texas, and Sun organizational cultures to produce an organization with deep points of cleavage, strongly influenced by political considerations. Some of the main problems of Corpoven seemed to be a board of directors with little or no internal rapport; a fragmented management team whose members belonged to very different organizational cultures; a sector of management politicized; and a complete lack of uniform systems and procedures. These were formidable obstacles, and in fact they could not be surmounted during the period. In 1981 the board was almost totally renovated, and some very talented executives were sent from Lagoven, Maraven, and PDVSA. The main problems have not yet been solved, however.

Against this background of organizational changes and conflicts, the operating companies had to embark upon an intense program of operational expansion. Considering the constraints just described, the results of the operational expansion during the 1976–1979 period were very satisfactory.

The Operational Expansion

The oil industry in 1976 was essentially extractive and semiobsolete. It had to be converted into a dynamic, modern, and expanding industry. In this the top leaders of industry were clear: new oil and gas reserves had to be found, the production capacity improved, refining installations remodeled, new international clients obtained, modern equipment bought, personnel recruited and trained for the future tasks facing the industry. The money-generating capacity of the industry had to be maintained. A new industry had to be created on the foundations of the old. This would take an enormous effort on the part of the oil employees and managers since they practically had to start from scratch. For the last 15 years the industry had lived mostly in the present without preparing itself adequately for the future.

The first task had to be the preparation of a set of policy guidelines, corporate objectives, and functional objectives—in short, a planning document. Petróleos de Venezuela, however, still lacked a planning group. Once again the industry was forced to rely on the operating companies. Of these only Maraven had long-range planning documents that could be utilized as the basis for an industrywide extrapolation. This company had already started on a planning cycle and was structuring a corporate planning organization. Fortunately a substantial amount of work had already been done at the functional levels of the main operating companies, and this work could now be integrated into an industrywide medium-range plan (5–6 years). Several professional societies such as the Venezuelan Geological Society, the Society of Petroleum Engineers, and the Society of Chemical and Refining Engineers had held, in the recent past, technical congresses and symposia in which high-quality papers dealing with exploration, production, and refining plans and strategies had been presented.[3]

On the basis of these documents a set of guidelines was prepared and discussed within the industry. The document was sent to the ministry, where it was further analyzed and approved, returning to the operating companies as the official policy guidelines for the industry.

Exploration goals were to search for light and medium oil reserves, both within the traditional areas and in new areas, including offshore; to conduct an exploration program for nonassociated-gas reservoirs with an emphasis on eastern Venezuela and the state of Guarico; and to evaluate the Orinoco heavy-oil deposits.

Production guidelines recommended maintaining oil production at the level of 2.2 million barrels per day (plus or minus 5 percent). Forty percent of this production was to consist of heavy oils. All necessary efforts were to be made to increase as soon as possible the production potential from 2.4 to 2.8 million barrels per day. Efforts in secondary recovery of oil were to increase even if economic analyses of some projects did not show the minimum accepted levels of corporate profitability (15 percent). Studies were to be conducted leading to the reactivation of marginal oil fields. And finally, all liquids were to be extracted from natural gas before its utilization for commercial purposes.

Refining guidelines advised initiating the studies and construction work leading to the conversion of the refineries from plants of high fuel-oil yields to plants of high distillate and gasoline yields. A supply of products to the domestic market was to be assured, and an increasing percentage of heavy oils was to be utilized as refinery feedstock.

Regarding the *domestic market,* modernization and expansion of service stations throughout the country was recommended. To improve the product transport system in the country, the building of the new terminals and pipelines would be required. And in general the nation should strive for a more rational consumption of hydrocarbons.

In *international trading and transport* efforts at diversification of clientele were to be intensified, albeit without the abandonment of traditional markets. Direct sales were to be emphasized. The utilization of traders or brokers was to be minimized especially within the western hemisphere. Sales strategy should be based on reliability of supply and optimizing the opportunities for the marketing of heavy oils. And last, but not least, obsolete units of the fleet must be replaced.

Materials guidelines were to increase purchasing of Venezuelan-made products, taking into account quality, prices, and delivery times, and to evaluate and assist local manufacturers of equipment to increase their capacity and quality control.

Regarding the *environment,* oil operations were to be conducted in such a manner as to protect the environment and quality of life.

Human resources goals included maintenance of a strict set of professional criteria for recruiting, training, career development, promotions, and remuneration. Manpower increases were to be planned in close cooperation with universities and other sources of trained human resources. Organizational analyses were recommended to improve the administrative structures and the managerial processes of the industry. And the best of relations with organized labor and professional societies were to be maintained.

To strengthen the *technological* capacity of the industry, INTEVEP, the Venezuelan Technological Institute for Oil and Petrochemicals, would be developed. Technological agreements signed with multinational oil companies were to be implemented in a timely fashion. Finally, the guidelines recommended that alternative sources of technology be evaluated continuously.

It is to the credit of Oil Minister Valentín Hernández that he recognized and accepted that much of the planning could only be done at the industry level, since the ministry was not organized to do this task.

Exploration

The creation of temporary task forces to evaluate the hydrocarbon prospects of the different areas of the country proved to be a very sensible move. These task forces were made up of four to six geologists and geophysicists from the different operating companies, coming together under a team leader. The groups had at their disposal all the information on the diverse areas contained in the files of the industry. For the first time in the history of oil exploration in Venezuela, the geological data and interpretations of all companies could be looked at in an integral manner. The groups had to report within 6 months to give PDVSA: the status of geological knowledge in each area, a recommendation for further studies, a ranking of the most important exploration prospects, and a proposal for an integral and systematic exploration program of the area. The main areas selected for immediate study were those already known to be of the greatest interest. Preliminary studies on the plausible oil and gas reserves in virgin areas of Venezuela had suggested that some 10 billion barrels of oil could be present in the still partially known or totally unknown sedimentary basins of Venezuela.[4] In descending order of amount of plausible reserves (in billion barrels) the most attractive unexplored prospect areas seemed to be as follows:

Gulf of Venezuela	4.0
Delta of the Orinoco River Area	4.0
Golfete de Coro	0.3
Gulf of Paria	0.3
South Lake Maracaibo	0.3
Offshore Puerto Cabello	0.2
Tuy-Cariaco Basin	0.3
Approximate total	9–10

Of course estimates could only be taken as the informed opinion of the geologist or geologists doing the work. Perhaps what was more important was the element of relative importance, or ranking, of each one of the areas.

The teams worked hard and in about 6 months, during late 1976 and early 1977, they came up with their recommendations. On the basis of these recommendations, three offshore rigs were contracted, two of them jack-ups (for shallow to intermediate water depths) and one self-propelled unit. One jack-up was operated by Maraven, one by Corpoven, and the third unit by Lagoven. These rigs were expensive to rent, in the range of $30,000 per day, in 1978 prices.

Maraven's Campaign. If geological ranking had been faithfully followed, Maraven would have started exploration drilling in the Gulf of Venezuela, the best-looking offshore area in western Venezuela. Because of a boundary dispute with Colombia which had lasted for many years and which had recently been reactivated, however, the Gulf of Venezuela was off-limits for exploration activities. Maraven was forced to go into the offshore Puerto Cabello area, Golfo Triste, where two broad anticlinal features had been mapped by seismic methods. Two wells drilled in these features did not find any significant show of hydrocarbons. A third well drilled to the west of the first two wildcats, in the offshore extension of a prominent land structure in East Falcon State, was also dry. These results confirmed the geologists' opinion that the young tertiary sediments in the offshore of East Falcon and central Venezuela were not important oil prospects.

Maraven moved into the Tuy-Cariaco basin, offshore from the State of Anzoategui. This basis was also a young tertiary trough mostly filled with Miocene and Pliocene sediments. It had an attractive, large uplift in the vicinity of the small island of La Tortuga. After a first dry well near the mainland, Maraven found modest amounts of good-quality (32° API) oil and gas in well MTC-IX, east of the La Tortuga island. In total, ten wells were drilled by Maraven in this basin up to 1982. In wells PMN-IX, EBC-2X, and MTC-4X, small oil and gas discoveries were made. At the end of this first round of exploration, Maraven had downgraded considerably its expectations in the offshore of Puerto Cabello and tended to concentrate future offshore exploration efforts in Tuy-Cariaco while waiting for the authorization to explore the Gulf of Venezuela.

Lagoven's Campaign. Lagoven concentrated efforts in the offshore of easternmost Venezuela, north of the Paria Peninsula and east of the Orinoco Delta. The Orinoco Delta drilling efforts proving largely unsuccessful, Lagoven directed its attention to the northern part of Paria Peninsula, west of the offshore Trinidad gas-bearing structures. The Patao-1 well found gas-bearing sands which tested about 90 million cubic feet per day, an excellent rate of production. Lagoven drilled eight more wells in this area in 1979 and 1980, finding a total of four gas "fields,"[5] and in 1981 it added two more discoveries of gas and some condensates.

Corpoven's Campaign. The drilling efforts of Corpoven in the offshore area were restricted to the Golfo de la Vela, north of the city of Coro. From 1978 to 1980 the company drilled five wells in this area. These were not real wildcats since eleven wells had been previously drilled nearby in the early 1970s. This campaign was very unsuccessful, to the point where the rig was transferred to Lagoven in 1981, ending Corpoven's offshore program.

Other Exploration Efforts. Although less spectacular than the offshore campaigns, the four operating companies conducted lake and inland exploration with reasonably good results. (See figure 8-1 for a map of their respective zones of activity.) Corpoven was very successful in eastern Venezuela and in some deeper Cretaceous wells in the center of Lake Maracaibo. Lagoven found new oil west of the lake, in the Perijá and Urdaneta areas. Maraven found excellent production in Motatan, southeast of the lake, and Meneven found new oil in the deeper Eocene reservoirs of the Ceuta area, in Lake Maracaibo.

In general exploration activity from 1976 to 1979 increased as shown in table 8-1. An overall evaluation of the first 4 years of postnationalization exploration efforts for light oil and gas would have to conclude that, although no great spectacular discoveries were made, significant new oil and gas resources were identified, probably of the order of some 2-3 billion barrels. The final figures will be known only when the new reservoirs are developed. In spite of these important contributions, it is now clear that the future of the Venezuelan hydrocarbons industry depends on the heavy-oil and, possibly, natural-gas deposits.

Production

Traditional Areas. Keeping production stable proved to be a formidable task. Venezuelan traditional producing areas were mature to senile. Some of the larger oil fields such as Lagunillas, Bachaquero, Tia Juana, and Oficina had been producing for more than 40 years; others such as Mene Grande for even a longer time. The newest Lake Maracaibo reservoirs were already 15 years old. Yearly rate of production decline was 25-30 percent. The degree of effort and cost of maintaining production at the level of 2.2 million barrels per day therefore increased continuously. Production expen-

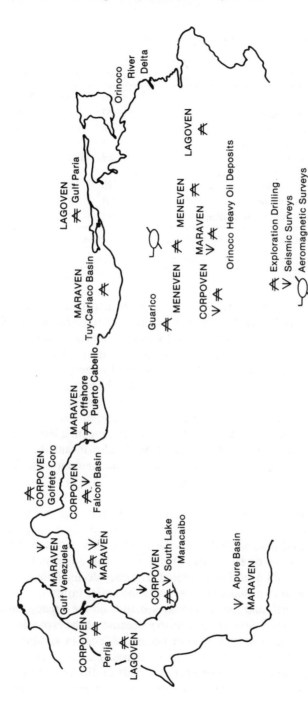

Figure 8-1. Exploration Activities by the Nationalized Oil Companies, 1976–1981

Table 8-1
Exploration Activity and Costs, 1976-1979

	1976	1977	1978	1979
Seismic lines (kilometers)	6,500	5,500	17,600	17,700
Exploration wells initiated	52	54	68	115
Exploration wells completed	58	49	61	118
Proven oil reserves by end of year (million barrels)	18,288	18,039	18,228	18,515
Proven gas reserves by end of year (million cubic meters)	1,180	1,185	1,211	1,249
Investments, (Bs millions)	403	413	778	1,380
Operational expenditures (Bs millions)	132	142	374	658

Source: Petróleos de Venezuela, *Informe Anual,* 1981 and other sources.

ditures (see table 8-2) increased from about $1.21 per barrel in 1976 to about $2.12 per barrel in 1979. This is easily understandable in light of the lower production yields per new wells and per newly repaired oil wells. In 1976 each new well drilled contributed an average of some 430 barrels per day. A repair in 1976 contributed an average of some 180 barrels per day; in 1979 this average had decreased to about 150 barrels per day.

Another negative factor was the increased time spent on the repair of wells. Whereas in 1976 a repair took an average of 6 days, the average in 1979 had increased to 9 days. Obviously more time on a less productive job meant a much more expensive operation. It also meant that more repair jobs had to be done in order to regain the same amount of production. In 1976 876 repairs were done; this figure rose to 1,470 in 1979—almost twice as much.

A logical question to ask is if the longer time per job and generally lower production yields were merely the result of increasing operational complexity and natural decline in the productivity of the reservoirs or if there was also loss of efficiency. This is a difficult question to answer, but there is little doubt overall operational efficiency declined somewhat as time went by due to two main factors: the substantial influx of younger, untrained personnel, and the additional demands on the already hard-pressed managers and specialists of the industry because of the expansion of activities. This loss of efficiency did not become evident in an abrupt manner. The oil industry works with great profit margins, a fact that tends to mask inefficiency. Further, it is extremely difficult to differentiate between normal reservoir decline and decline in operational efficiency. Because of these

Table 8-2
Venezuelan Production Data, 1976-1979

	1976	1977	1978	1979
Production wells completed	254	347	584	668
Production wells repaired	876	916	1,079	1,470
Drilling rigs/year	33	44	58	93
Investments (Bs millions)	873	1,470	2,178	3,124
Operational expenditures (Bs millions)	2,322	2,835	3,259	3,883
Production gained (thousand barrels per day)	500	502	510	540
Production cost per barrel ($/barrel)	1.21	1.48	1.92	2.12
Productivity per well drilled (barrels per day)	430	390	350	310
Time spent on repairs (days per well)	6	7	8	9
Production potential (thousand barrels per day)				
Light/medium	1,879	1,733	1,789	1,631
Heavy	743	759	751	884
Totals	2,622	2,492	2,540	2,515

Source: Petróleos de Venezuela, *Informes Anuales,* and unpublished documents.

factors, there was no clear perception of this situation for the first four years of nationalized operation.

Personnel involved in production activities increased considerably, both as directly hired employees and as contracted workforce. If we assume that about 60 percent of the total work force directly hired by the industry is engaged in production work, we can conclude that industry labor force in the production area rose from approximately 13,000 employees in 1976 to 18,000 employees in 1979. This increase was largely the result of increased well-drilling and well-repair.

Secondary recovery projects increased significantly. In 1978 Maraven completed its M-6 steam-injection project in the Tia Juana field, east of Lake Maracaibo. The largest steam-injection operation in the world, this project is designed to recover about 120 million barrels of heavy oil which would have otherwise been left behind, in the subsurface.

The greatest accomplishment of the oil industry during the period 1976-1979 was to keep production at the level of about 2.2 million barrels per day. To increase the production potential to 2.8 million barrels per day proved to be impossible, however. In fact, if anything was apparent, it was that production potential had dropped, going from about 2.6 million barrels per day in 1976 to 2.5 in 1979. Of all operating companies, the only one which actually increased potential was Maraven, going from about 590,000 barrels per day in 1976 to some 630,000 per day in 1979. Lagoven kept its potential at around 1.1 million barrels per day, and both Meneven and Corpoven experienced reductions. It became clear that if Venezuela wanted to increase production potential significantly, this could not be done in the traditional areas, but it would have to be done by developing any offshore new fields and/or the Orinoco heavy-oil deposits.

The Orinoco Belt of Heavy-Oil Deposits

PDVSA started work in the Orinoco Belt as soon as it obtained administrative control of the project, creating a coordinating office and establishing a master plan to conduct an evaluation of the entire area. The evaluation had the following objectives:

To conduct general exploration so as to obtain a better knowledge of the distribution and quality of the crude oils in the area

To select the most prospective areas and evaluate them in detail

To carry out production and secondary recovery pilot projects in sites previously chosen

To evaluate the upgrading technologies available with a view to utilize the most promising in the development of the Orinoco heavy oils

To set up a production development program in order to obtain about 600,000 barrels per day of oil in 1995 and 1 million barrels per day of oil in 2000.

Because of the great extension of the area to be evaluated, some 700 kilometers in an east-west direction and about 60 kilometers from north to south, it was decided to divide it in four portions and to assign a portion to each one of the operating companies, from east to west: Lagoven, Meneven, Maraven, and Corpoven (see figure 8-2).

Cerro Negro, Lagoven. Each one of the areas had its own geological and topographical characteristics. The easternmost, assigned to Lagoven, is called Cerro Negro and is located to the south of existing oil fields (Jobo, Morichal, Pilon, Uracoa). Heavy-oil reservoirs had already been identified in Cerro Negro through the early work conducted by the ministry group. The objectives of Lagoven consist of developing, upgrading, and transporting this crude oil. The target set is 125,000 barrels per day by the year 1988 and about 500,000 barrels per day by the year 2000.

The production stage includes the drilling of producing wells and steam-injection wells. Two pilot projects, one in Jobo and the other in

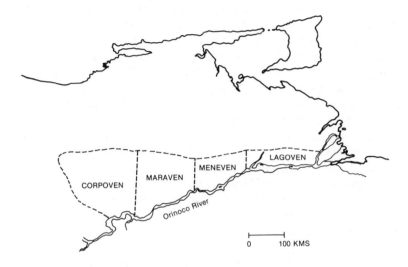

Figure 8-2. Orinoco Heavy-Oil-Deposit Areas Assigned to Each
Operating Company

Cerro Negro proper, have been built to evaluate reservoir behavior, optimum rates of production and injection, and the economic aspects of the
operation.

To accomplish the target of 125,000 barrels per day will require the
drilling of about 1,000 wells. The most expensive aspect of the operation,
however, will be the crude-oil upgrading plant, consisting of two installations in parallel for the distillation, retarded coking and hydrogenizing of
the crudes. The upgrading mechanism will center on improving the
hydrogen-deficient nature of the Orinoco crudes to increase the hydrogen-
to-carbon ratio. The chosen process is that of retarded coking, long in use
and already proven in commercial scale.

Planning the execution of this project started in 1978 and was practically completed by 1981. Lagoven has spent over 1 million man-hours in
planning activities. The company stuctured a project team composed of
Lagoven and contract personnel (mostly from Bechtel) to prepare the
master plan and budget estimates required to obtain shareholders' approval.

Main Elements of Lagoven's Master Plan. The project includes the drilling of about 1,000 wells and the construction of surface installations,

steam-injection plants; the upgrading complex, powered by an electric plant of 300 megawatts; a loading terminal; a pipeline and a small camp consisting of offices, housing, roads, aqueduct, and other facilities.[7] The plan estimates that about 3,000 persons will work permanently in the area.

Lagoven will utilize as far as possible the facilities already existing in the neighboring towns such as Temblador and in Ciudad Guayana rather than embarking on large-scale construction of new facilities. Special care will be taken to harmonize oil activities with the agricultural and cattle-raising development already existing in the area.

Environmental considerations are of primary importance, since the activity of the upgrading and steam-injection plants will generate polluting emissions and the handling of large volumes of crude oil will present significant risks of contamination of rivers. The master plan includes environmental studies as well as provisions for continuous monitoring of the environmental impact of the project, to correct in time any damage to the ecological system.

The basic organization for both the planning and construction stages was as follows: The main responsibility and authority were with the board of Lagoven. Reporting directly to the board was a task force of about eighty Lagoven staff members supported by a planning advisor, a company to supply the Lagoven organization with all the human and technical requirements needed to plan the project. The planning advisor would report to the Lagoven project team. The construction stage would require a general coordinator for execution (GCE), a company that would coordinate the execution of the different components of the project, obtain the equipment and materials required, execute those portions of the project specifically assigned to it, act on behalf of Lagoven as quality-control agent in the manufacture of equipment and execution of construction, and train personnel.

The general coordinator for execution would have to be a company characterized by experience in petroleum megaprojects, expertise in multiple disciplines, large-scale construction technology, and advanced control systems. These characteristics, according to Lagoven, excluded the possibility of selecting a Venezuelan company. It would have to be a foreign company associated with a Venezuelan firm or firms.

The contract with GCE would be a service contract; payment would consist of reimbursable costs plus a stipend for services. We will return in more detail to the organization selected by Lagoven for this important project, as it became a major political and emotional issue during 1980 and 1981.

Guanipa, Meneven. The area immediately west of Cerro Negro, called Guanipa, was assigned to Meneven for excellent reasons. It is just south of Meneven's traditional oil fields and is essentially a geological continuation of those fields. The objective of Meneven is to produce about 100,000 barrels of heavy oil per day from this area by 1988. This new production will be diluted with lighter oils to obtain a blend of 17 °API oil, which can be refined in the Meneven installations at Puerto La Cruz.

Meneven's plan for the evaluation and development of the area includes about 6,000 kilometers of new seismic lines; the drilling of about 300 exploration wells to define the areas of better hydrocarbon content; the drilling of 1,300–1,800 production wells; the construction of all associated production facilities, including 100 production stations and an electric plant of 50 megawatts; and eventually, the construction of steam-injection plants to supplement the primary recovery of the oil. Although this project is considerably less complex than Lagoven's it will still require a considerable investment, estimated at some Bs 6 billion (at 1980 prices), as well as some 1,000 new employees.

The Evaluation Work of Maraven and Corpoven. In addition to the production projects of Lagoven and Meneven, there are two other evaluation programs under way in the Orinoco area. Maraven has been assigned the Zuata zone, west of Meneven's project, to conduct exploration and general geological and reservoir evaluation. In order to do this, the company has already drilled more than 180 exploration wells and executed over 2,000 kilometers of modern seismic lines. Maraven has found essentially two types of oil accumulation in its area: heavier oils of 8–10° API in thick sands (60–70 meters and less heavy oil, of 15–20° API in thinner sands south of the town of Pariaguán. Maraven plans to conduct steam-injection projects in carefully selected areas.

Corpoven has activities in the westernmost portion of the Orinoco area, which borders on the El Baúl uplift. The area, called Machete, probably has poorer prospects than the others because the reservoir sands tend to be thinner. After some 50 exploration wells, the company has found heavy-oil deposits of still uncertain value.

Evaluation and Development Results until Mid-1980. The work conducted in the Orinoco area by the operating companies of Petróleos de Venezuela has been an improvement over the earlier, less systematic one conducted by the ministry group. Since 1978, when actual work standard, to the end of 1981, about 11,200 kilometers of modern seismic lines had been done and more than 600 exploration wells drilled with a very high rate of success. Table 8-3 summarizes the work done up to 1977 and the work done from 1977 to the end of 1981.

Table 8–3
Activity in the Orinoco Heavy-Oil Belt

	Up to 1977	1977 to End 1981
Detailed plans	None	Yes, almost completed.
Seismic lines (kilometers)	12,800	11,200
Exploration wells	110	603
Aeromagnetic surveys	— —	A complete one.
Production tests	No Data	420
Pilot projects of steam injection	— —	2, Jobo and Cerro Negro
Reserves identified (million barrels)	No Data	3,600

Source: Based on Petróleos de Venezuela, *Informe Anual,* 1981.

Refining

The Venezuelan refining pattern was characterized by a large percentage (about 62 percent) of residual-fuel-oil yields. This had been a logical result of the petroleum policies pursued by Venezuela's main client, the United States, during the last 30 years, a policy favoring both the import of foreign light crude oils to be refined domestically in order to produce gasolines and distillates, and of low-cost residual fuels to be used as heating oil. In the last 10 years, the U.S. market for residual fuel oils had clearly been weakening because of more demanding environmental measures and a strong trend for conservation. The use of domestic coal had increased in place of fuel oil in those areas of the United States where environmental considerations were less rigidly applied. Combined with the change in the U.S. market there also was a change in the Venezuelan local market characterized by a very rapid, almost explosive growth in the consumption of gasolines and diesel fuels a growth due to the lack of mass transport systems and to the low prices of these products in that market. There was little question, therefore, that the Venezuelan refineries had to change their configuration in order to keep Venezuela competitive in the international markets for refined products and be able to supply to the local markets increasing volumes of gasolines and distillates. A plan was devised to accomplish this in a systematic fashion. The four major refineries in the country: Amuay (Lagoven), Cardón (Maraven), El Palito (Corpoven), and Puerto La Cruz (Meneven), were subject to careful analyses in order to find the optimum way to convert them into plants yielding substantially larger volumes of light ends. Amuay developed a plan calling for the installation of new coking, catalytic cracking, and alkylation plants. The coker used Exxon technology and basically converted crude oils into light ends and a very heavy, practically solid coke-type residue. The Cardón refinery project consists of two stages and includes revamping of the existing catalytic cracker (completed in 1981), a new hydrocracking unit, an experimental demetallization unit, and desasphaltizing and alkylation plants.

The El Palito project included a new catalytic cracking unit together with vacuum distillation and alkylation plants. Finally, the Puerto La Cruz refinery project included a new catalytic cracker and an alkylation unit together with a major revamping of the existing atmospheric distillation units.

The chronology of these major projects was to be as follows:

1981   El Palito (inaugurated in June 1981)

1982   Amuay

1983   First stage at Cardón

1985   Puerto La Cruz

1986   Second stage at Cardón (hydrocracker)

The net result of these projects will be to permit Venezuelan refineries to produce up to 65 percent of gasolines and distillates and to accept much heavier crude-oil intakes than has been the case up to now. By completing the projects, the Venezuelan oil industry will be able to add about 200,000 barrels per day of gasolines to the 190,000 barrels per day produced at the time of nationalization and about 60,000 barrels per day of gas oils. At the same time residual fuel oil production would decrease by more than 90,000 barrels per day.

By the end of 1982 the status of these projects was as follows:

El Palito   Completed in June 1982

Amuay   About 90 percent completed

Cardón   First stage under construction, slightly delayed

Puerto La Cruz   Undergoing bidding for construction

The total investments estimated for these projects are of the order of Bs 15,000 million (1981 prices).

These projects came under attack by the political left and from some members of the technical staff of the Ministry of Energy and Mines. Opposition also came from members of the board of the Venezuelan Engineering Society, a professional group which has become increasingly politicized since 1960 and from the former oil minister, Dr. Juan Pablo Pérez Alfonzo.

The arguments against the refining projects of the industry can be summarized as follows:[8]

> Huge investments should not be dedicated to expanding the domestic refining capacity and to subsidizing the irrational domestic consumption of gasoline.

> The technologies utilized in the revamping of some of the refineries were chosen to benefit multinational enterprises, since some of them were still in the experimental stage.

> The flexicoker was not a convenient choice for Amuay, according to work done by independent consultants.

> The contracting of the works at the El Palito refinery had been irregularly conducted.

> It would have been perhaps preferable to import gasolines than to embark on these costly projects.

Some of these arguments were strictly political; few had real economical or technical bases. The first one, from the socialist MAS party, was made on the false assumption that refining capacity of plants would increase. This was not the case. Overall refining capacity would remain essentially stable; what would change would be the percentage of products. There was much merit in the claim that these huge investments would encourage waste, especially if additional volumes of gasolines and distillates continued to be sold at the same low prices. In fact an indispensable complement to these investments—if they were to be really justified—should be an increase of the domestic price of refined products. The claim was still valid in 1982, since only the gasolines had increased in price.

The second argument charged the Venezuelan oil industry with using only those refining technologies which would benefit the multinational oil companies, adding that these technologies were still in the experimental stage. In fact, with the exception of the Amuay flexicoker and the frankly experimental demetallization plant to be built by Maraven, all other units were pretty standard technology. The flexicoker to be used in Amuay by Lagoven was an Exxon patent being used commercially since 1978 in Japan, in a plant which had been running for three years without any major problems. The demetallization unit to be constructed by Maraven in Cardón was an experimental, 2,400-ton-per-day plant, to be built after extensive bench-scale tests carried out in the Shell Laboratories in Amsterdam had shown that very heavy crude oils with high sulfur and metal content could be converted into lighter, deeply desulfurized and demetallized oils after being processed with the help of special catalysts. Both of these units had been selected after much analysis by industry task forces consisting of refining, chemical, and mechanical engineers from the operating companies as well as from PDVSA.

The third argument had its origin among members of the technical staff at the Ministry of Energy and Mines who had been viscerally opposed to the flexicoker from the very beginning. This staff had in fact commissioned a study of the Lagoven project to a consulting firm which concluded that the Lagoven project was not the most convenient, basically because of the difficulties which would be encountered in disposing of the cokelike residue.[9]

The fourth argument had to do with the contracting of the works at the El Palito refinery, where in fact some irregularities were committed. After receiving the bids from companies trying to get the contract, the management of the operating company (Llanoven) decided, rather unwisely, to let a European firm ranked in the fifth place to join the four leading companies on the short list, in order to pressure those companies into bringing their prices down. The predictable result of this maneuver was that the firm presented such a low bid that it should have been given the job, except that it just did not seem to have the desirable technical qualifications. After much analysis the bidding process was repeated and the contract finally went to a different firm, but not before the original low bidder had protested loudly. This protest was picked up by the political opposition, which now had a real case of poor managing to complain about.

The argument in favor of importing gasolines rather than investing in new refinery facilities was made by Dr. Pérez Alfonzo. Its basis was economic, although Pérez Alfonzo did not dwell on details. The decision to invest in domestic refinery facilities, however, was not so simply an economic one but had an important strategic ingredient. It was thought that Venezuela should not depend on imports of a vital product, without which national transport would come to a standstill. Venezuelans also believed that the investments made in the refineries would stimulate the local economy during the construction phase of the plants as well as later on, something which could not be said of imports. The poor economics of refining for the domestic market was largely a consequence of the low prices of products in that market. If these prices were increased, the overall economic attractiveness of the new refinery facilities would also increase, although, admittedly, the projects would probably never have a satisfactory rentability.

The Local Market

Two problems of particular importance in this field were: (1) the rapid growth of consumption of refined products, which was starting to cut deeply into the volumes earmarked for export, and (2) the obsolescence and inadequacy of the transport and sales infrastructure—service stations, terminals, pipelines, and storage.

The first problem was the consequence of the very low, subsidized prices imposed by democratic governments for the last 25 years and of a pattern of public transport characterized by a total absence of mass transport systems and the proliferation of individual cars. The increase in consumption can best be illustrated by the statistics listed in table 8–4. Gasoline consumption more than doubled during the 1970s and that of diesel oils almost tripled. Total consumption in the local market represented only 5 percent of total national production in 1970 but jumped to about 17 percent of the total national production in 1980.

The obsolescence of plant and equipment serving the local market was easy to explain. The oil companies had refrained from investing in a sector which was markedly unprofitable due to the artificially low prices of refined products in the country. In fact the cost to the companies of gasoline and other products being sold in the local market clearly exceeded the sales prices. The oil industry estimated a direct subsidy to the local market of about Bs 600 million in 1977. The indirect subsidy was much greater, the opportunity cost of the products which could have been exported at prices up to eight times higher than the internal, regulated prices.

The efforts of the Venezuelan oil industry to solve the problems just outlined have so far been unsuccessful. The local market remains one of the most acute problem areas of the Venezuelan oil industry. This is true for several reasons. The vagueness of policies concerning the sector has inhibited a clear, determined drive for improvement; Pricing policies, in particular, have almost entirely been dictated by political considerations. The control of service stations throughout the country is no longer in the hands of the oil industry, but in individual, private hands due to a political decision made by President Pérez in 1975. This makes quality control of products and services impossible to conduct.

The managerial quality in the local market organizations is low. The local market had been increasingly taken over by CVP during the first half of the 1970s and, as a result, the local market divisions in the oil companies had been largely dismantled. The better managers had either been transferred to other activities or had asked to be transferred. Much of the remaining staff was close to retirement or had low potential for career advancement. The overall effect of this situation was a downgrading of the activity. Companies started looking at the local market as an area where investments, either in capital or in people, should not be made. Although there was a reversal of this attitude after nationalization, it has not yet been possible to upgrade the sector to the rank of the other, more profitable activities.

Table 8-4
Increase in Domestic Consumption of Hydrocarbons, 1970–1980
(million barrels per year)

	1970	1971	1972	1973	1974	1975	1976	1977	1978	1979	1980
Gasolines	25.7	27.3	29.5	31.9	34.5	38.4	42.8	47.9	52.2	55.6	58.5
Diesels	9.1	10.3	11.2	13.8	14.9	15.7	17.2	17.5	22.0	24.8	28.6
Other products	38.2	39.6	42.7	46.6	41.3	34.7	33.8	33.8	34.9	39.7	45.3
Totals	73.0	77.2	83.4	92.3	90.7	88.8	93.8	99.2	109.1	120.1	132.4

The local market has intensive interaction with outside private groups such as the Federation of Gasoline Distributors, the Federation of Service Station Owners (FENEGAS), and the Association of Candle Manufacturers, who buy paraffin from the industry. This interaction has been frequently marked by conflict and, in the case of the candle manufacturers, by allegations of corruption and favoritism on both sides.

The local market is that sector of the industry where government intervention has been the greatest, not only in the area of quality control but also in the field of policy and strategy formulation. There is a local market division in the Ministry of Energy and Mines, which in the last few years has been very active trying to establish some programs, such as the use of liquefied petroleum gas in automobiles, which should be handled exclusively by the industry. This interface has often resulted in significant friction and increased inefficiency.

International Trading and Transport

The volumes of crude oil and petroleum products exported by Venezuela during the period 1975–1980 decreased sharply but their geographical distribution remained essentially constant, as shown in table 8–5. The increase in exports to Europe was almost entirely due to the Italian market, which went from 9 million barrels in 1975 to 40 million barrels in 1980, mainly on the strength of contracts signed by Maraven with ENEL, the Italian state electrical company.

What changed significantly was the identity of the buyers. Up to the time of nationalization, most of the Venezuelan exports were sold through the

Table 8–5
Venezuela's Main Customers for Exports of Heavy Oil and Petroleum Products, 1975 and 1980

	1975 Percentage	1980 Percentage
United States	33	25
Curaçao and Aruba	24	25
Canada	12	9
Central America	13	12
Europe	10	19
Other countries	7	10
Total Exports (million barrels)	760	680

Source: "Petróleo y Otros Datos Estadísticos," Ministry of Energy and Mines, Caracas, 1980, p. 86.

marketing facilities of the multinational oil companies. With the exception of Shell of Venezuela, the rest of the oil companies based in Venezuela disposed of their oil through their holding companies. Shell had created, in the early 1960s, a trading department which directly sold increasing volumes of hydrocarbons. Still, at the time of nationalization, about 75 percent of Venezuelan exports were directly marketed by the multinationals. In 1980 this proportion was of only 40 percent and, what was more important, the trading organizations of the operating companies had been substantially strengthened.

Also significant was the change in the nature of exports. In 1975 only 25 percent of crude oil exports were heavy (less than 22° API). In 1980 heavy oils made up 50 percent of total crude-oil exports. This new export pattern reflected the changing nature of Venezuelan oil reserves and the increasing use of light oils in the domestic market. It was made possible by the marketing skills of the trading organizations, since some of the Venezuelan heavy oils were of much lower quality than the oils being marketed by Middle Eastern and African producers. Of special importance in this effort were the aggressive strategies of Maraven and Lagoven in the market of heavy-oil specialties, such as asphalt and lubricants, and in the U.S. market of heavy crudes and residual fuel oils. About 70 percent of Maraven's crude-oil exports in 1981 were heavy oils, while more than 60 percent of its exported products were residual fuel oils. In the same year, about 90 percent of all Lagoven's exports consisted of heavy oils and products.[10]

The Trading Organizations. One of the real innovations of the nationalized oil industry was the creation of organizations to sell oil in the international market. In this field, once again, Petróleos de Venezuela had to rely heavily on the experience of some of the operating companies.

Maraven had by far the most experience. Under Alberto Quirós and Carlos Castillo this company had created, since 1967, a very solid group of international traders. The organization of Maraven was basically as shown in figure 8-3. The supply department served as the connecting link between the production and refining functions and the trading function. Its main responsibility was to create the availability of crude oils and products that the trading organization would need to satisfy the requirements of potential clients. The supply department was also responsible for keeping trading informed of the existence of crude oils and products for which buyers should be found. A sale usually started with an inquiry from a potential buyer. This inquiry would be handled by specialists both in the geographic area where it originated and in the types of crude oils or products required. The reply of the company would usually be in the form of an offer, quoting price and conditions of payment and delivery. The planning department was essentially concerned with market analysis and prediction. This group

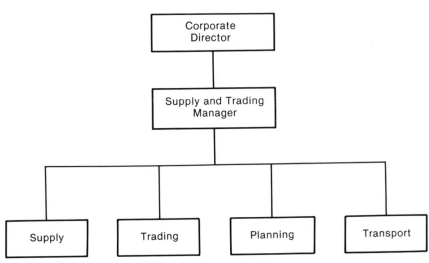

Figure 8-3. Organization of the Trading Department of Maraven

would estimate demand in the short and medium terms and guide production and refining efforts of the company as well as the establishment of marketing strategies.

Although the organization existing in Maraven was used as a general guide in the consolidation of the trading organizations of the other companies, something else had to be added. Up to 1976 every company had developed its own marketing strategies since they had different owners. After nationalization, marketing strategies had to be developed for the whole industry. This work was done by two high-level marketing committees. The first of these committees was composed of the trading managers of the operating companies and the coordinator of trading in PDVSA and his supporting staff. It had the task of recommending pricing and marketing strategies for the industry. The second one, composed of the trading directors, the presidents of the operating companies, and the PDVSA director in charge of trading, would analyze the work done by the first, essentially technical committee and would add the political, longer-term considerations. The resulting recommendations would be presented to the board of PDVSA and, eventually, to the minister of Energy and Mines for approval.

With more than one Venezuelan oil company selling oil abroad there were obvious risks, the greatest of which was to have more than one company competing for the same client and selling oil at less than an optimum price. Several cases of this type occurred during the first years of nationalization, especially during the period in which the companies with new trading organizations were eager to obtain a share of the market. This situation was satisfactorily resolved when the technical trading committee started to meet to analyze clients and potential markets and to assign them to each of the companies. The assignments were based on the type of hydrocarbons being marketed, the experience of the companies in different geographical areas and the existing commercial ties between the client and the organizations.

Human Resources

The number of personnel in the Venezuelan oil industry decreased consistently from 45,600 employees in 1957, the year in which the last oil concessions were granted, to 23,088 in 1975. However, by 1979, employment had again increased to about 30,000 employees. There were many reasons for this postnationalization increase:

The intensive exploration program which had begun in 1977

The new production projects in the fields of secondary recovery and in the utilization of natural gas

The increasing activity in production drilling and well-repair work which was required to maintain the Venezuelan production potential at about 2.5 million barrels per day

The evaluation of the Orinoco heavy-oil deposits

The remodeling and new plant construction started in the refineries in 1977

The new projects being undertaken by the local market sector

The development of INTEVEP, research institute for the petroleum and petrochemical industries

The absorption of Pequiven, the petrochemical group of companies

The growth of the holding company and the creation of commercial intelligence units abroad.

Considering the significant expansion in activities of the industry, the employment increase during the 1975–1979 period cannot be considered abnormally large. Of more concern has been the significant decrease in the average experience of the oil employee. In 1975 about 23 percent of the employees had less than 5 year's experience, but in 1979 43 percent of the employees belonged to that group and fully 57 percent had less than 9 year's experience. This was bound to have a negative effect on productivity. The new employees could not do as much work as the more experienced and the more experienced now had to dedicate considerable time to the training and guidance of their new working companions. As General Alfonzo Ravard expressed in his inaugural speech to the Seventh Technical Petroleum Seminar, held in Caracas on June 27, 1977: "The petroleum engineers have still a third responsibility in addition to operating the industry and doing research. They have to serve as teachers and advisers for the new engineers who are now joining the ranks of industry."[11] There can be little doubt that this factor has played a role in the cost increases experienced by the industry since nationalization. However, most of the cost increases can be explained by the significant increment of activity as well as by inflation.

During this period the industry embarked on a very serious effort to standardize personnel norms and procedures. Prior to nationalization, companies had done their recruiting, training, and personnel development in accordance with the plans of their holding companies and had followed their own procedures. Now, under a single ownership, they could no longer openly compete for the same human resources or treat them in a radically different way once employed. The personnel specialists of the operating companies formed task forces to agree on norms to be applied by everyone. Starting salaries, intercompany transfers, job evaluations, benefit plans were progressively standardized and, although some differences remain, there are no longer any great procedural differences in the treatment of personnel in the different companies. The differences still existing are derived from the corporate style of the individual organizations and are not likely to disappear in the short term.

Technology

Evolution of the Technological Agreements. Most of the technological support required by the industry immediately after nationalization came from the multinational oil companies, through technological agreements negotiated in 1975. These agreements covered most industry activities from exploration to refining. In 1976 there were twelve different agreements, but intense negotiations with multinationals and the search for other sources of technological support paved the way for the reduction of these agreements to four by the end of 1979.

The agreements had obvious advantages. They contributed to a smooth operation of the industry by supplying support to day-to-day activities, and more important, to long-range projects. They put at the service of the operating companies abut 300 specialists in the different areas of activity of the industry. These specialists not only did their own work but substantially contributed to the training of nationals. An important technological liaison with the outside world was provided and a means to compare the level of efficiency of the Venezuelan oil-industry operations with similar operations carried out by multinationals abroad.

The agreements also had certain disadvantages. An effective transfer of technology was not guaranteed. In most cases this was not the fault of the multinational companies but of the lack of Venezuelan personnel qualified to act as recipients for the transfer. Much of the support was received without making a real effort on the receiving end to learn "how to do it next time." The wide range of support tended to inhibit creative thinking on the part of the local staff. Why think hard if the solution was at hand, sometimes just by making a telephone call?

The mechanisms for payment were indefensible. A fee based on production was strongly reminiscent of the royalty payments under the no-longer-extant concession program.

The day-to-day operational pressure was usually too great to permit the organizations a careful administration and evaluation of the use being given to the agreements. According to J. Villalba, Maraven was the company which had done the best job in this respect.[12]

The high quality of the support received tended to discourage the search for alternate technological sources, something which had to be done in order to minimize a dangerous dependence on a few sources. In the search for new sources, Lagoven was probably the most successful since very solid contacts were made with large engineering companies such as Bechtel, Parsons, and Williams Brothers. Some of these contacts eventually resulted in agreements which substantially contributed to the diversification of technological support.

Performance of the Agreements. In order to illustrate how the technological agreements worked it is useful to summarize the results of a survey conducted by Maraven in early 1979 (see table 8-6). This survey dealt with the three basic questions: How dependent is Maraven on outside technological support? What has been the quality of Shell's technological support? When will INTEVEP's contribution become significant?

Most of the employees replying "as well" or "almost as well" to the question "How would Maraven do without outside technological support?" worked in exploration, production, and refining operations. The less optimistic staff were engaged in long-range planning and projects. Some of the conclusions of the Maraven survey are interesting. There was a high degree of satisfaction among the staff with the functioning of the agreement. Most employees felt, however, that the cost was too high. Most employees agreed that it was difficult to put a concrete figure on the value of the services obtained by Maraven through the agreement. Not all services included in the agreement seemed to be really required. The two main reasons to justify

Table 8–6
Results of Maraven's Survey on Technological Agreements, 1979

Degree of Company Dependence on Outside Technological Support

	Percentage of Total Opinion	
	Activities that do not require outside technological support	*Activities that do require outside technological support*
Materials	20	80
Exploration and production, long-range studies	25	75
Exploration and production operations	50	50
Refining	30	70
Computing	60	40
Average percent	37	63

Quality of Shell's Technological Support

	Percent of Opinions			
	Excellent	*Good*	*Regular*	*Poor*
Materials	35	60	5	—
Exploration and production, long-range studies	28	55	18	8
Exploration and production, operations	20	45	25	10
Refining	10	55	35	20
Computing	8	58	20	14

When Will INTEVEP Generate Enough Technology?

	Percent of Total Opinions
In 2–5 years	8
In 5–8 years	20
In 8–10 years	30
In over 10 years	42

How Would Maraven Do without Outside Technological Support?

	Percent of Total Opinions
As well	10
Almost as well	20
About half as well	65
Very poorly	5

the existence of the agreement were the shortage of human resources in the company and the intimate knowledge Shell and Maraven had of each other. Although most technological transfer had gone from Shell to Maraven, the existence of the agreement also allowed a certain degree of reverse technological flow (at no cost to Shell). The exchange of technological information derived from the agreements among operating companies was very limited, even after the confidentiality clause was removed.

The Renegotiation of the Agreements. In January 30, 1979 the first meeting of the steering committee on technology took place in PDVSA, chaired by Vice president Arreaza, with the following persons attending: PDVSA's Directors Domínguez and Reimpell; President Quirós and myself, a director, for Maraven; President Rodriguez Eraso, Vice-president Sugar and J. Trinkunas for Lagoven; Vice-president Romero for Meneven; Vice-president Calderon Berti for INTEVEP. R. Irving of PDVSA acted as secretary. In this meeting the committee agreed on the following points:

1. The committee will establish the guidelines for negotiating the new technological agreements. Once established, these guidelines will be consulted with the minister of Energy and Mines before being put into effect. The negotiations being carried out by the operating companies will be carefully monitored.
2. The operating companies will establish their needs for technological support no later than February 30, 1979.
3. The committee will base the negotiation strategies of the industry on the contents of Lagoven's paper dated October 13, 1978 and will structure an organization as proposed in Maraven's paper dated October 23, 1978.
4. A technical committee will be created.
5. Each operating company will form its own negotiating team.[13]

The document submitted by Lagoven recommended three basic objectives: an effective transfer of technology, free sharing of technological support received by any company with the other companies of PDVSA, and payment based on man-hours of services required.

The document submitted by Maraven dealt essentially with the organiztion the oil industry should adopt to carry out the negotiations. Maraven's paper distinguished three distinct levels of involvement in the negotiations: a coordinating level at the holding company, a corporate level in each one of the operating companies, and a functional level in each main area of activity. The coordinating level would supply general strategic guidance, making sure that all agreements shared an identical conceptual basis and incorporated the basic objectives that industry was trying to obtain. The corporate level would conduct the negotiations on the basis of the objectives coming from the coordinating committee and would be supported by the technical data prepared at the functional levels.

After many working sessions of the groups just described, a set of guidelines for the negotiations were finally agreed upon by PDVSA and sent to the Ministry of Energy and Mines. The minister sent back a letter to industry, in November 1979, approving the guidelines. In the letter the minister said, "We have studied the general guidelines proposed by PDVSA for the renegotiation of the technological agreements which are deemed necessary and agree to them."[14]

The letter from the minister added the wish of government to include, as fundamental ingredients of the new contracts, the already well known concepts of a real transfer of technology, payment only for services rendered, the nonconfidential nature of the technological support, and an ideal maximum duration of the agreements of 2 years.

The Mechanisms of Negotiations. By this time negotiations had already been in progress for some time. It had been decided to start negotiations between Maraven and Shell. This was desirable for three main reasons. First, excellent rapport and great mutual respect existed between the Maraven and Shell management teams, unlike the more formal and distant relations existing between Lagoven and Exxon. Second, the crude oil available to Shell at the moment was in short supply, which probably made this company more inclined to a rapid agreement on the technological area than Exxon, which had a much stronger oil supply position. And third, there was a feeling that whatever was agreed between Maraven and Shell would also be agreed by the other parties.[15] Accordingly, Maraven presented Shell with a request for technological support which included 40,000 man-hours per year of advice in the different areas of activity of the company; training of Venezuelan nationals in Venezuela and abroad; about eighty-five specialists on full-time assignment and a miscellaneous package of computing software, technical documents, and several other services. For all these services Maraven offered a payment of $25 million per year. Shell replied that the alternate value of such a package to Shell, what Shell could obtain by using these resources in any area of the world, would be $75 million. These were the starting positions. After several weeks of intense negotiation, Shell had moved down to about $50 million and Maraven had moved up to about $30 million. When the possibility for further progress had dwindled to practically zero, Maraven suggested that a meeting be held with the board of Shell in London. This idea was accepted; the meeting took place in November 1979, and was attended, on behalf of Maraven, by the company's president A. Quirós, A. Volkenborn, J. Zemella, and me. Shell's officers in attendance included a member of the Board of managing directors and several of his high level staff. The following is an unofficial record of that meeting:[16]

> Shell stated its interest in staying in Venezuela under a long-term agreement. Shell was still worried about the pending business between Shell and the Venezuelan government (tax claims and guarantee fund) although it admitted that Maraven had little or no say in these matters. Shell also admitted its interest in agreeing to a technological support contract and in obtaining renewed access to Venezuelan crude and products.
>
> Maraven presented Shell with a summary of the many different agreements, commercial and technological, existing between the two companies and emphasized the importance of a global approach to these agreements.
>
> Shell said that it understood the technological needs of Maraven perfectly and that such needs had a cost of about $75 million, mentioning again the concept of "alternate value." It added that, if Maraven had been a Shell affiliate, the net cost of the package would be of about $40 million.
>
> Maraven said that the relationship between Maraven and Shell had always been characterizied by stability and by a trusting, long-term relationship. On this basis, Maraven was prepared to keep supplying Shell with hydrocarbons on a long-term basis, although these hydrocarbons also had a considerably higher "alternate value" in the world markets. Therefore, the position of Maraven in the supply field deserved a quid pro quo in the technological field.
>
> The meeting ended with both Shell and Maraven agreeing that a solution should be forthcoming in the next few weeks.

As a result of this meeting, Shell and Maraven went on to agree, in February 1980, to a new technical agreement with the following characteristics:

50,000 man-hours of services per year (10,000 man-hours more than had originally been requested by Maraven)

Eighty-five Shell specialists working for Maraven in Venezuela

Material procurement services for up to $85 million per year

Training and cross-postings of Venezuelan staff by the Shell organizations

No price increases during the life of the contract.

As a result of these negotiations, agreements were also reached with Exxon and Gulf with the general results shown in table 8–7. The total cost for industry of the renegotiated agreements was $123 million per year, a reduction of almost 40 percent over the costs of the agreements which expired in December 1979.

Table 8-7
Results of Renegotiations of Technical Agreements

	Lagoven–Exxon	Maraven–Shell	Meneven–Gulf
Assigned staff (at cost), in Venezuela	105	85	50
Man-hours of services per year			
Basic	80,000	50,000	30,000
Optional	—	10,000	3,000
Maximum material procurement			
($ millions)	150	85	50
Training and cross-postings	Yes	Yes	Yes
Price escalation with inflation	Yes	No	No
Total cost ($ millions)	64	38	21

INTEVEP. During the period 1976–1979 the research and development institute for the petroleum and petrochemical sectors underwent consolidation. Excellent headquarters were designed[17] and built in Los Teques, 15 miles southwest of Caracas, and intense recruiting of mostly young scientists took place. The group of researchers from the hydrocarbon division of IVIC (the Venezuelan Institute of Scientific Investigations) was transferred to INTEVEP.

The most important early task for the institute was to define its objectives. In several meetings held with INTEVEP's board the top management of PDVSA stressed that INTEVEP should not become an institute for technical services but a true institute for research. As such it had to resist the temptation to engage in short-term activities which pertained to the field of operations. This is the general orientation the institute has taken. Its work is being done in several areas such as seismic data processing, palynology, and nannoplankton, for exploration; core analysis, compaction and subsidence studies, heavy-oil production and secondary recovery, and drilling engineering, for production; and catalysts for demetallization of heavy oils, boilers and steam equipment, and lube oils, for refining. Of some concern has been the frequent change in leadership. During its first 6 years of activity the institute had already had three presidents: Martorano, Vásquez, and Segnini. Top management has had, in general, a very rapid turnover, and finance and administration still are very weak aspects of the organization.

An Overall Appraisal of the Operational Expansion

The expansion of activities which took place in the Venezuelan oil industry during the period 1975–1979 was very significant. During these years, the industry made intense efforts to find new hydrocarbon reserves, to increase production potential, to modernize its refineries, to put order into the local market, to obtain new international clients for its exports, and to create a new generation of managers, technicians, and workers. In this endeavor the industry was only partially successful. Hydrocarbon reserves identified were mostly gas and heavy oils but not light oils. The production potential did not increase. The local market kept deteriorating. On the other hand most of the work done on the refineries has been well executed and, once fully finished, it will give the industry much more flexibility to operate at varying levels of refinery output than it has been the case up to now; secondary recovery projects have been continued; the huge Orinoco heavy-oil projects are under way; significant skill has been shown in the acquisition of new clients in the international market, and many of the younger employees have shown that, given the right example to follow, they can absorb the good work habits of the old timers.

Expansion is dangerous, of course. If not carefully controlled and monitored, it can lead to chaos. The most dangerous aspect of expansion is work force growth. The labor force in the industry grew by almost 40 percent during the period. This posed the problem of absorbing enormous volumes of new employees into the system at a fast rate. In many cases the result has been only partial absorption, distortion of the system by the introduction of new group values, and relaxation of discipline and loss of morale. New employees have not been properly assimilated by the organizations, a situation which might well lead to inefficiency.

At the same time expansion has also meant increased expenditure. From repairing about 870 wells in 1976, the industry went on to repair almost 1,500 wells in 1979. Drilling of exploration and production wells more than doubled in the same period. To get this job done involved much more rig activity and contracting. The only way to be able to secure the equipment needed was to give the operators much more authority and to relax the norms and procedures which had been used for years. This led to waste and in some instances to corruption. There have been known cases of company employees accepting bribes from contractors in the form of new cars, trips abroad, even houses. Nepotism has been known to occur, contracts being handed out to close relatives. Although not widespread, such cases have had a very damaging effect on the organizations mostly because the culprits have rarely been punished. In some cases the culprits have been rewarded with promotions, which have further served to demoralize the honest sectors of the organizations.

In spite of these deviations, expansion was generally well conducted. It was also inevitable. The industry simply had no choice but to grow.

Notes

1. In 1978 Maraven had about forty-three top executives assigned to other organizations within the oil industry. In 1981 eleven of the thirty-six top officers of PDVSA had come from Maraven, seven from Lagoven, three from Meneven, two from the ministry, one from Corpoven, and the rest from the outside.

2. Verbal communication from a Lagoven official.

3. First Venezuelan Seminar on Geology, Mining and Petroleum, Maracaibo, 1975; Second Venezuelan Refining Seminar, Puerto La Cruz, 1976; First Symposium on Extra-Heavy Oils, Maracay, 1976; Fifth Technical Seminar on Oil, Caracas, 1977.

4. G. Coronel, "Recursos de Hidrocarburos en las areas inexploradas de Venezuela," *SVIP Bulletin,* Caracas, August 1972.

5. Lagoven, Resumen de Actividades, 1981.

6. *Resumen* 391, "La Faja Petrolifera del Orinoco," 1980.

7. Ibid., p. 23.

8. Summarized from the communique of MAS about the oil nationalization published in *El Nacional,* D–17, of January 17, 1978; the document on oil, of the Venezuelan Engineering Society, Puerto Ordaz, May 1978 and the lecture given by Dr. J.P. Pérez Alfonzo in his home, "Camurana," in October 1978.

9. The consulting firm was Bonner and Moore. A copy of the study was immediately given to the Venezuelan press by ministry officials and was amply used by the political left to back their views with a more scientific basis.

10. Maraven, *Informe Anual,* 1981, and Lagoven, *Informe Anual,* 1981.

11. "Cinco Años de Normalidad Operativa," Petróleos de Venezuela, Caracas, 1981, p. 127.

12. Julián Villalba, unpublished Ph.D. dissertation, Massachusetts Institute of Technology, Cambridge, Mass., 1982.

13. Minutes of the meeting were handed out to all attending.

14. My private files.

15. This is not unusual in negotiations concerning multinational oil companies and involving several of "the sisters." In 1960 during the Cuban–U.S. government quarrel over the refining of Russian crude oil in Cuba's refineries and the subsequent refusal of the American companies to do so, Shell had to go along with the decision made by its American sisters. For an account of this incident, see P. Odell and L. Vallenilla, "The Pressures of Oil" (London: Harper & Row, 1978).

16. From my private files. Note that in November 1979, spot market prices in Rotterdam were well on their way to $40 per barrel.

17. By an American firm of architects, W. Pereira, of Los Angeles.

9

The Financial Performance of the Nationalized Oil Industry

The financial performance of the nationalized Venezuelan oil industry has been mostly successful. Income per barrel has grown steadily, as has National Participation. (National Participation is defined as all income taxes, including royalties, plus Petróleos de Venezuela's participation plus the net profits of the operating companies.) Operating costs have also increased, but they still represent a similar and sometimes lower percentage of total income per barrel than in prenationalization years. Table 9–1 shows these indices.

The Evolution of Investments

Prior to nationalization, investments had remained relatively low, at the level of $1 billion per year in the period 1970–1975 (about $250 million). Table 9–2 shows the change in investment levels which took place with nationalization. The table illustrates the importance of the production sector. After nationalization, however, production investments came to represent a somewhat smaller percentage of total investments as other areas of activity started to receive considerable attention.

The rapid increase in investment levels was a logical consequence of nationalization. Multinational companies could not afford to invest in projects which would only start providing returns near or beyond the year in which most concessions would expire, 1983. With the state as owner, the oil industry could not embark upon long-range capital projects. The first year after nationalization was a period of transition. Investments remained at the same level of prenationalization years, while the industry concentrated in planning the new capital expenditures which would be required. In 1977 investment levels doubled, and in 1978 doubled again. Exploration and production investments represented 75 percent of total capital expenditures in 1977. Exploration drilling of the deeper prospects of Lake Maracaibo and Eastern Venezuela (more than 16,000 feet deep) accounted for 75 percent of total exploration investments. About half of total production investments

Table 9-1

Financial Performance Indices for the Venezuelan Oil Industry, 1970-1980

Year	Total Income (Bs/barrels)	National Participation (Bs/barrels)	Total Costs Bs/barrels)[a]	National Participation (Percent of Total Income)	Operating Costs (Percent of Total Income)
1970	8.36	4.52	3.57	53	22
1971	10.46	5.70	3.06		
1972	11.32	6.90	3.38		
1973	15.84	10.33	3.40		
1974	41.69	35.36	4.08		
1975	42.95	33.78	6.12	79	15
1976	45.22	38.30	6.07		
1977	49.37	41.28	6.18		
1978	49.24	38.68	9.66		
1979	70.33	54.18	11.45		
1980	104.03	86.27	17.42	85	14
1981	110.00	90.00	20.00	81	15

Source: 1970-1978 based on "Petróleo y Otros Datos Estadísticos," Ministry of Energy and Mines, Caracas, 1980, 1981, estimated from PDVSA *Informe Anual* (Annual Report), 1981. (U.S $ = Bs 4.3).
[a]Includes depreciation.

were connected with the drilling of wells, but 22 percent of total production investments were already associated with long-range projects.

In 1978 exploration investments amounted to Bs 780 million, almost 20 percent of total capital expenditures. Production investments, again, were largely associated with drilling and included the acquisition of sixteen new land-based drilling rigs. Almost 30 percent of total production investments were dedicated to long-range projects. Investments in the refining sector increased by a factor of 4. Three large projects were started: the remodeling of the Cardon refinery catalytic cracking unit by Maraven, the construction of a new catalytic cracker in El Palito refinery by Corpoven, and the construction of the Amuay refinery flexicoker by Lagoven. The Cardon refinery project was undertaken by Lummus, the El Palito project by Foster Wheeler, and the Amuay project by Fluor. Together, they made up a major part of a plan to change the product mix of the Venezuelan refineries, allowing them to refine heavier crudes and to yield ligher products.

In 1979 exploration investments again increased, to Bs 1,400 million, about 25 percent of total capital expenditures. Offshore exploration drilling accounted for most of the expenditure but exploration drilling in the

Table 9-2
Investment in the Venezuelan Oil Industry, 1970-1979
(million Bs; U.S.$ = Bs 4.3)

Year	Exploration, Production	Transport	Refining	Other	Total	Exploration, Production as Percentage of Total Investment
1970	943	107	216	28	1294	73
1971	945	134	170	38	1287	73
1972	758	67	131	44	1000	76
1973	829	201	47	44	1121	74
1974	1506	163	113	178	1960	77
1975	661	198	144	207	1210	54
1976	768	128	112	105	1113	69
1977	1583	127	185	214	2109	75
1978	2429	476	713	382	4000	61
1979	3607	455	1383	395	5840	62

Source: "Petróleo y Otros Datos Estadísticos," 1980.

Orinoco heavy-oil areas started in earnest. Production investments continued to be largely associated with drilling and almost 800 new wells were started during the year. Projects already made up almost 35 percent of total production investments. Refining capital expenditures doubled to about Bs 1,383 billion.[1] The large projects started in 1978 made significant progress and the revamping of the Cardon refinery catalytic cracking unit was completed. Four new tankers were delivered to the Lagoven and Maraven fleets to replace aging equipment.

The investment figures shown in table 9-3 for 1980 and 1981 have been taken from the PDVSA Annual Report for 1981. The increase in the investment levels was again dramatic. Capital expenditures more than doubled in the period 1979-1981. The largest portion of this huge increase was represented by four main areas of activity: exploration, the Orinoco heavy-oil projects, the refinery projects, and the acquisition of five new tankers, plus the ordering of eight more units to bring total tonnage of the oil fleet to about 730 million tons by 1983.

When compared with 1976 investments, 1981 levels were higher by a factor of about 13. This brings up the question of how the industry has managed such a rapid expansion. It must be remembered that for 15 to 20 years prior to nationalization, the prevailing atmosphere of the industry had been one of contraction. By 1975 expenditures had been cut to the bone and

Table 9–3
Investment in the Venezuelan Oil Industry, 1980 and 1981
(Bs million)

	1980	1981
Exploration	2,194	2,696
Production	4,127	6,464
Refining	2,952	2,604
Others	488	1,757
Totals	9,761	13,521

manpower was down to about half of what it was in 1957. This atmosphere of chronic contraction had created an attitude in managers and employees characterized by austerity, caution, and reluctance to change. In order to expand investments by a factor of 13 in 5 years, an abrupt change had to take place in the attitude of the personnel of the industry. This was not the least of the challenges faced by industry and, although it could not be quantified, there is no doubt that it came very high on the list of worries of top oil-industry executives.[2] For many years the oil industry's favorite password had been "Save." Now it had become "Spend" or, even more difficult, "Spend wisely."

The transformation of the management philosophy of industry from one of big savings into one of big spending had obvious complications. To save very often means not doing, whereas spending always means doing. Once managers grow used to not doing, it becomes very difficult for them to start making big and rapid decisions. The commitment to spend imposes new pressures on the organizations. It was not enough for the nationalized oil industry to have the money and the best of intentions in order to be able to spend. The Venezuelan oil industry found that it was much more difficult to execute the budget than to get it approved. From 1978 to 1981 the average percentage of execution of the production budget of the two largest operating companies was about 90 percent. But the other two companies executed only about 70–80 percent of their budgets during the period 1976–1978. Budget execution had to do with quality of the organization, good planning, efficient contracting and materials procurement systems and, above all, sound management.

The Planning and Budget Cycle

Perhaps the most useful tool in the efficient execution of the industry's plans and budgets was the planning and budget cycle. From the very beginning the coordination of finance of Petróleos de Venezuela had been in very capable hands, first with Pablo Reimpell, later with Oliver Campbell. These men had been conversant, in their previous organization, with a systematic cycle of medium-term and short-term planning documents which guided the implementation of the yearly budgets. This cycle was now adopted by PDVSA and by each operating company and was divided, essentially, into four parts: a conceptual stage, in which planning scenarios would be drawn and corporate objectives and stategies formulated; a first-draft stage of the yearly budget, totally in line with the objectives defined previously (in the second quarter of the year); the final-draft stage of the budget, both capital and operating (in the third quarter); and finally presentation to shareholder (PDVSA) and approval (in the fourth quarter). This was a companywide exercise involving the corporate planning and finance groups as the coordinating entities as well as the participation of all functions.

The detailed sequence was similar in all of the operating companies, as follows:

Conceptual Stage

February 7	Revision of the results of the preceding year.
February 15	Results of the preceding year presented to PDVSA
February 28	Establishment of short- and long-term scenarios. Revision of corporate objectives. Meeting with the board.
March	Establishment of functional objectives and approval by the board. Preparation of planning and budget guidelines to be distributed to the functions and operating divisions.

First-Draft Stage

April 1	Estimate of human, technical, and financial resources to implement the conceptual guidelines.
April 15	Meetings between operational management teams and staff of main office to establish common grounds for the budget.
May 8	Revision of the results of the first quarter of the current year.
May 13	Presentation of the results of the first quarter of the current year to PDVSA.
May 19	Coordination between operational plans and manpower requirements, the basis for the operating budget.

June 2–13	Putting together the first draft of next year's budget.
June 16–24	Discussion of this draft by the board of directors.
June 25–26	Revision of the first draft.
June 27	Board of directors meets to discuss revised first draft.
June 30	First draft is sent to management of all functions and operating divisions.

Final Draft Stage

July	Detailed functional budgets are prepared.
August	First-semester results of current year are presented to PDVSA.
August 25	Capital and operating budgets now ready, in the hands of the finance group.
September 1–15	Presentations on the budgets made to the board.
October 1–6	Budgets in book form are delivered to the finance group of PDVSA.
October 21	Capital Budget is presented to the board of PDVSA.
November 10	Detailed revision of the operating budget.
November 25	Operating budget is presented to the board of PDVSA.
December 6	Shareholder meeting at PDVSA approves budgets.

This exercise had several advantages. It made all members of the organizations very conscious that a budget was not simply a list of things to do and to be prepared once a year but an integral part of a larger, long-range planning exercise. It served to ascertain that no individual projects would be carried out unless they fitted into corporate and industrywide objectives. It could be used by the holding company, PDVSA, to make sure that there were no duplications of efforts in the industry and to select only the best projects presented by each company. And it could be used by the operating companies to monitor on a continuous basis the execution of their budgets.

The Problem of Increasing Operating Costs

Although operating costs in 1975 and 1981 represented a similar percentage (about 15 percent) of the total income per barrel, the absolute figure was about three times greater in 1980 than in 1975. In a presentation made to the members of the Energy and Mines Committee of the Venezuelan National Congress, the minister of Energy and Mines, H. Calderón Berti, said that the increase in operating costs in the period 1976–1980 had four main ingredients: inflation, the effect of the government law passed in December 1979 increasing salaries of all government employees by an average of 20 percent, the results of the labor collective contract signed with the oil worker's union in 1980, and the significant increase in oil-industry activity.[3]

According to the figures presented by Calderón Berti the increase in operational costs during the period 1976–1980 could be explained as follows: 40 percent of increase could be attributed to inflation; 30 percent to the labor contract; 15 percent to the new government salaries increase; and 15 percent to the increased activity.

Inflation was an external factor over which the oil industry had no control. The same thing could be said of the government's decision to increase the salaries of their employees and, to a certain extent, of the results of the labor collective contract signed with the oil workers' union. The only cost element substantially controlled by the industry was the level of activity. But, again, production volumes and income were intimately tied to the level of activity. It has already been mentioned that to produce the same amount of barrels per day required ever-increasing effort, time, and hence money on the part of the industry. When operating cost increases became noticeable by the political sector, however, the industry became subject to considerable scrutiny and much of the unconditional trust characteristic of the first four years of nationalization was replaced by caution or even suspicion.

Oil Income and Rising Public Expenditure

The growing distrust of the political world was reinforced by the mounting national economic problems. During the first big oil "crisis," 1973–74, Venezuelan income had suddenly quadrupled. The national budget went from Bs 14 billion in 1973 to Bs 42 billion in 1974. President Carlos Andrés Pérez found himself with total majority in Congress, great popularity and extraordinary amounts of money. As Mauricio García Araujo tells us:

> In the short year period from 1972 to 1975, gross domestic product at current prices doubled; government revenue tripled; foreign exchange inflow quadrupled; and the level of international reserves of the Central Bank quintupled.[4]

The Pérez government embarked upon an ambitious program of economic development investing huge sums of money in megaprojects such as steel mills, hydropower plants, shipyards, and social-interest programs

such as a massive scholarship plan and massive agricultural loan program.[6]
As Pedro Pick says in his analysis of state-owned industry in Venezuela:

> The results of this well-intended super plan were inflation of 24 percent per
> annum, deficit in the balance of payments, huge national debt, an inflow
> of illegal immigration, further spreading of corruption, a labor force that
> had high turnover and low productivity and further deterioration in quality
> of social serviced and distribution of wealth.[6]

What Pick describes is a classic example of an economy distorted by oil
revenues. Imports grew very rapidly, at the rate of 30 percent per year during the period 1974–1978. The results of the investments made by the
government failing to materialize, government spending and borrowing
went out of control. In 1977 the government of Carlos Andrés Pérez realized that it was heading into financial chaos and started trying to put the
brakes on what was defined as an overheated economy. The money supply
was curtailed, imports became subject to increasing control. The objectives
of these actions were to curb inflation, to stimulate the construction sector,
and to reduce imports. By the time Pérez handed over the government to his
successor, Luis Herrera Campíns, March 1979, none of these objectives had
been attained. President Herrera claimed, in his inaugural speech, that he
received a mortgaged country. Garcia Araujo writes, "The new government
. . . promised to control public spending and cool an overheated economy
through conservative fiscal and monetary policies."[7] In doing so, Garcia
Araujo adds, the new government "plunged the economy into recession."
In addition, the Herrera government undertook a program of price
deregulation and eliminated most of the subsidies for consumer products.
These measures produced significant and abrupt price increases. García
Araujo tells us that "the cost of living index for the city of Caracas rose 11
percent"[8] during the last quarter of 1979. This led the government to decree
a general increase in salaries and wages which further stimulated infla-
tionary pressures and which had an important role to play in the increase in
operating costs in the oil industry in 1979 and 1980. The economy of
Venezuela was characterized during 1980 and 1981 by little or no growth
and by a pronounced loss of the investors' confidence int he ability of the
administration to lead the country into recovery. In 1982 the bottom fell out
of oil prices, and the already shaky economic situation became critical.

It was in this increasingly deteriorating economic atmosphere that the
oil industry had to operate during the period 1977–1981. It became quite
evident that oil income might no longer be sufficient to satisfy the ever-
rising fiscal expenditures. The oil industry estimated tht, even assuming increases in export prices of 13 to 15 percent per year and increases in the local
market prices to bring them to 50 percent of export prices and even assuming that the country could develop a production potential of 2.8 million

barrels per day, the country could still develop serious problems of balance of payments and fiscal deficits during the 1990s. The studies made by the industry led to the conclusion that the consumption of oil in the Venezuelan domestic market will be in the range of 800,000 to 1,800,000 barrels per day by the year 2000, depending on whether a slow rate or a rapid rate of economic growth are assumed (0.9 percent per year or 3.7 percent per year of increase in gross domestic product) It was also concluded that to reach a production potential of 2.8 million barrels per day, the new areas, especially the offshore and the Orinoco heavy-oil belt should contribute about 1.5 million barrels per day by the end of the century. Of this volume, the Orinoco area should contribute about 1 million barrels per day.[9]

There was no doubt, therefore, that the development of the Orinoco heavy-oil deposits was indispensable to maintain the oil industry capable of generating the income needed by the country. At the same time, however, this development would require huge investments which had to be diverted from other sectors of the economy. This posed a real dilemma for the country.

The Oil-Industry Fund

From the inception of nationalization, the president of Petróleos de Venezuela, General Rafaél Alfonzo Ravard gave much attention to the implementation of a mechanism which could ensure the self-financing of the industry. The law of nationalization contained the provision that 10 percent of the net value of oil exports should be given to PDVSA to be used in the financing of the oil industry projects. In time, this fund grew with the addition of the net profits of the operating companies until it reached, by 1981, a sum close to $8 billion. It will be apparent that the existence of such an important potential source of monetary resources was very much in the minds of Venezuelan political leaders. With the mounting national economic problems, it is not surprising that members of government started to look to the oil fund as a possible solution to their already chronic money problems. By early 1982 the fund had not yet been used for this purpose, thanks to the strong stance of Acción Democrática in the defense of the self-financing mechanisms of the industry. Gonzalo Barrios, that party's president, publicly accused the finance and energy and mines ministers of Luis Herrera's cabinet[10] of asking Acción Democrática for their support to utilize Bs 2.5 billion of the oil fund to pay some government debts for which bank loans had proven impossible to obtain.[11] Barrios said that Acción Democrática felt that "it was highly dangerous for the country to utilize financial resources of the oil industry to cover the waste and inefficiency of government entities in debt." There was no longer much doubt, however, that the government would keep trying to tap the PDVSA fund for more general, probably less-rewarding purposes.[12]

Notes

1. According to "Petróleos y Otros Datos Estadísticos," Caracas, 1980, p. 124. Discrepancies are quite frequent between financial figures as given in this ministry publication, and the PDVSA annual reports. Although they are almost never significant and can eventually be reconciled, they can be confusing.

2. Average investment per employee is not a particularly useful index to illustrate this concept, which has to do with behavior rather than with numbers. However, for the sake of illustration, in 1976 investments per employee were of about Bs 60, while in 1981, the equivalent figure was Bs 322.

3. *El Nacional,* April 23, 1981, p. D-1.

4. Mauricio Garcia Araujo, "Venezuela's Economic Challenge for the 1980s," a talk delivered at the Venezuelan–American Association, January 28, 1982, New York.

5. The Fundación Mariscal de Ayacucho provided scholarships for about 10,000 Venezuelan students both at home and abroad; the program still exists and has had only partial success. Many of the agricultural loans went to people who knew little or nothing about agriculture and used the money for several other purposes.

6. Pedro Pick, "Letter from Cambridge: Proposal for Reorganization of Venezuelan State-Owned Industry," Center for International Affairs, Harvard University, 1982.

7. Garcia Araujo, "Venezuela's Economic Challenge for the 1980s," p. 4.

8. Ibid., p. 6.

9. *Resumen* 391, "La Faja Petrolífera del Orinoco," p. 15.

10. *El Nacional,* June 15, 1982, D-11.

11. A jumbo loan of $2.5 billion being negotiated with a group of European banks, including English banks, had fallen through because of the political tension developing between Venezuela and the United Kingdom over the Malvinas Islands crisis.

12. This finally took place in September 1982 (see Chapter 12).

10 The End of the Honeymoon, 1979 and 1980

Relationship with External Sectors, 1976–1979

In June 1976 the president of Petróleos de Venezuela, General Rafaél Alfonzo Ravard, was the speaker at the annual journalists' dinner. There he explained the success of the nationalized oil industry in the first semester of activities as due to

> the solidarity and support that (we) have received from the national government and from all economic, labor, and political sectors, a support which has been indispensable to the successful execution of our tasks. . . . Not less important has been the understanding shown by the members of the press and other communication media. This understanding has been basic . . . to our success.[1]

These words accurately described the predominant nature of the relationship between the national oil industry and the rest of the country during the period 1976–1979. It should be remembered that the basic underlying reason for nationalization had been a desire for national self-assertion. Venezuelans had to prove to themselves that they could manage the complex oil industry as successfully as the foreign enterprises had. They wanted the nationalized industry to succeed so that they could no longer feel inferior. As time went by and the oil industry kept operating normally and obtaining satisfactory profits, there was widespread satisfaction and pride in the Venezuelan public regarding this performance.

From the outset, however, there were important foci of resistance and systematic criticism against PDVSA and the operating companies. One was at the Ministry of Energy and Mines and was led by high officers who had exercised considerable control over the industry during the period 1970–1975 and had now seen this control substantially decreased with the creation of PDVSA. Resistance from the ministry was expressed in several ways. A task force was created in 1977 to write a new hydrocarbons law, which would have restored all lost power to the ministry. Frequent delays occurred in giving technical approval to drilling, production, or refining projects not because there were good technical reasons to do so but because the officers at the ministry would not agree with the choice of sites, with the contractor selected, or with the quality of the managerial decision. Another important focus of criticism was the political left, led by the socialist party MAS, which had always argued that the top managers of the nationalized oil industry were mere stooges of the multinational oil companies, a claim which never made significant inroads into public opinion. MAS's criticism of the oil industry often lacked a minimum of technical basis and seemed to be fundamentally inspired in ideological considerations. MAS was never happy with the way nationalization had taken place. It had been "too orderly, too pacific." Negotiations had prevailed over conflict. Technological and trading agreements had been signed with the multinational oil companies. All of this, according to MAS, made for a make-believe type of nationalization without fundamental changes. The Venezuelan oil industry, MAS claimed, still was very much in the hands of the international oil companies.

Still another systematic flow of criticism came from politically independent, usually strongly nationalistic individuals. The most influential of these individuals was Juan Pablo Pérez Alfonzo, the former oil minister, who also tended to see the nationalization of the oil industry as an unfinished process. He defined it as *"chucuta,"* Venezuelan slang for a docktailed animal, usually a dog. Pérez Alfonzo shared the MAS perception of the oil nationalization as a process cut short. Periodically he convened meetings at his Caracas home, where he would receive his guests seated at the head of a ping-pong table, wearing a baseball cap and a sports shirt and would lecture on basic Venezuelan problems, usually related to oil.[2] The strongest attack on the nationalized oil industry made by Pérez Alfonzo came in October 1978, when he delivered a public lecture criticizing the following aspects of its activities:

Financial Results. According to Pérez Alfonzo, the national income derived from oil in 1977 was "no higher than in 1973" (p. 12 of Pérez Alfonzo's mimeographed lecture notes).[3] This claim was wrong on all counts since both total income and, more important, per-barrel income were much higher in 1977 than in 1973.

Production Decline. Pérez Alfonzo complained, and he suggested that such a drop was due to technical negligence on the part of the industry (p. 17). The reality was that Venezuelan reservoirs were in a very advanced stage of depletion. It was no longer possible to reverse the declining trend of their production capacity.

Oil Reserves. Pérez Alfonzo claimed that the official figures of oil reserves given by the oil industry were produced "in bad faith" and by "incompetent'' personnel (p. 25). In fact, Venezuelan oil reserves were calculated according to very strict, totally professional criteria by industry reservoir engineers and then certified by ministry personnel.

Oil Prices. Pérez Alfonzo observed that the average price of Venezuelan oil exports was lower than the reference price for the Arabian light marker crude (pp. 10, 11, 23). This was true and inevitable, as the average barrel exported by Venezuela was much heavier, had a much lower API gravity than the 34° API of the Arabian marker crude, and consequently also had a lower commercial value.

Exploration. Pérez Alfonzo strongly criticized the decline in both production potential and the exploration program of PDVSA. He said that exploration was ''subject to the whim of the successors of the multinationals, who did not risk anything (personal) in the venture'' and added, ''It is hard to accept that bureaucrats and technocrats should have the authority to gamble with the money of the Venezuelan people'' (p. 30). Of course, there was no better, known, way other than exploration to increase oil reserves and hence production potential.

Investments. Pérez Alfonzo worried about the investment programs of the industry, calling them ''fantasiosas'' (fantasies) and objected to the proposed investment of Bs 80 billion for the 1980s. Pérez Alfonzo's worries deserved great sympathy since Venezuela had embarked upon very costly megaprojects during the presidency of Carlos Andrés Pérez. To continue spending at these enormous levels suggested a high probability of waste. However, the Venezuelan oil industry seemed to have no choice but to invest in a number of activities designed to find new reserves, modernize the refineries and develop the Orinoco heavy-oil deposits.

Technical Support Agreements. In his criticism of the oil industry, Pérez Alfonzo was especially harsh on the technical agreements existing between the nationalized oil companies and the multinational corporations (p. 36). Pérez Alfonzo claimed that ''all Venezuelans have condemned those agreements,'' but his claim seemed to be grossly exaggerated. If this had really had been the case, the agreements probably would have never been signed.

In general terms, Pérez Alfonzo had become by 1978 the strongest critic of the nationalized oil industry. He used exceptionally harsh language to refer to Petróleos de Venezuela, their executives and plans. On page 69 of the mimeographed transcript of his lecture, he described the oil executives as ''guided by their selfish tribal interests,'' as ''the tools of powerful foreign interests,'' and as ''vain technocrats blinded by their power.'' On page 42 he made an extremely interesting statement: ''We never had the need to nationalize the oil and iron concessions," interesting because he had been perhaps the greatest advocate of the nationalization of those basic industries for the last 30 years.

In order to understand Pérez Alfonzo's position it is essential to know more about the circumstances of his life during the period. The man was very seriously ill. He had cancer and knew his time was limited.[4] He had always been a highly disciplined and austere Venezuelan and was now anguished about the collective madness of his countrymen, about the way they had wasted the immense riches obtained from the liquidation of a nonrenewable resource, about the seemingly incurable frivolity of Venezuelan society and leaders. His time was running out, and all he had been preaching for many years seemed to have fallen on deaf ears. He became a radical, using increasingly strong words to make his case. People who had very extreme views came close to him during those days and contributed greatly to the general tone of his lectures and press releases. In a way Pérez Alfonzo seemed to become—in his last months of life—the ideological prisoner of a small group who possessed a visceral distrust of top oil-industry management.

An influential group which also had some strong words against the national oil industry was the Venezuelan Engineering Society. The society allowed, in May 1979, the publication of a document prepared by some of its members, in which several activities of the oil industry came under heavy attack.[5] The strongest attack was directed against the top management of the industry and had to do not with their professional competence, but with their patriotism. The industry should be managed, the document claimed, by "men who did not have a past which would force them to accept situations contrary to the welfare of (our) country." It added that "the role of the multinational oil companies in the contracting of the projects of the industry should be a passive (advisory) one and not a deciding one as it has been up to now."

In the document it was argued that the board of Petróleos de Venezuela should include a representative of the Engineering Society. This was an unusual request, especially since the board of PDVSA at that time included seven engineers. None of these engineers acted as a representative of the society, nor could they. If all professional sectors in the country had made and been granted a similar request, the board of PDVSA would have required dozens of members.

Another request that seemed hard to grant was that "representatives of the Engineering Society must be present at all the bidding processes for the construction of projects in the oil industry." The operating oil companies had their bidding norms and procedures and could not alter them to allow the presence of outsiders who could conceivably have vested interests in the results of the biddings.

It was claimed, inaccurately, that "The technological dependence (on the multinational oil companies) has increased after nationalization."

The Engineering Society urged that "All newly graduated petroleum engineers be employed by the oil industry. Even if they do nothing useful, employing them would only increase production costs by some $0.03 per barrel, whereas the industry is now paying $0.40 per barrel in technological assistance." This statement contained several inaccuracies. The industry was paying a net average of $0.15 per barrel for a very wide range of technological services, which included the full-time activities of about 300 foreign specialists of long experience. The quality of this support could not be compared to the "help" that industry would be able to obtain from new graduates. Even more worrying, however, was the opinion advanced by the authors of the document that all new graduates should be automatically employed. If adopted, such a policy would have caused, in time, an inevitable drop in the average quality of employees and in the efficiency of the industry.

The statement that "INTEVEP should develop a capability for exploration in order to enable (this company) to maintain a satisfactory exploration effort" suggested a fundamental lack of information on the industry and the roles that each company of the oil sector should play. INTEVEP was a research and development company, not an operating company.

Finally, the Engineering Society argued, "The existence of Petróleos de Venezuela can be justified only if a functional scheme is created for the operating companies: a drilling company, a production company, a refining company." The idea that operating companies should not be integrated but of a functional character had been carefully considered and discarded in 1974. The big arguments against it were that such an idea would eliminate integrated organizations able to plan on a corporate basis and would convert existing profit centers into cost centers.

In spite of criticism such as the Engineering Society's, the Venezuelan oil industry had a veritable honeymoon with the country for about 4 years. In 1977 President Pérez could say with pride: "The Venezuelans have proven to themselves and to the world that (they) are capable of controlling (their) basic industries without resorting to international confrontation, within the law and acting with total responsibility." He added, "Now we know that we are capable. We can say that we have freed ourselves of the (inferiority) complex induced by more powerful nations to make us believe in the need for their tutelage."[6]

Pérez shared this perception with most of the country. The major political parties, Acción Democrática and the Christian Democrat COPEI, seemed happy enough to receive information on the oil industry in regular congressional subcommittee meetings or in courtesy visits paid to the offices and installations of the companies. From 1976 to 1979 there were very few congressional hearings on the issue of petroleum. During this period, the president of PDVSA, General Alfonzo held monthly meetings with President Pérez, a sign of deference and support which could not go unnoticed in the country, but one which did not particularly please the Energy and Mines minister, Valentín Hernández. Although the personal relations between Hernández and Alfonzo were never good, they never reached the point of open warfare. Hernández was always most supportive of the industry, even when this meant deep disagreements with his own staff at the ministry, a staff generally hostile to PDVSA. After some major battles, such as the ones for the control of the Orinoco heavy-oil project and for the construction of the flexicoker in the Lagoven Amuay refinery had been won by PDVSA, the staff at the Ministry became reasonably restrained, although the frictions never disappeared.

From Total Respect to Partial Interference

During 1979 the Iranian revolution and then war between Iran and Iraq reduced the oil available in world markets between 4 and 6 million barrels per day, a volume no other oil-exporting country could totally replace. As a result world oil prices started climbing very rapidly after some years of comparative stability. By December 1979 the prices in the world market had reached $36 per barrel and, in the Rotterdam spot market, some oil changed hands at $40 per barrel.[7]

Although Petróleos de Venezuela did not sell oil in the spot market, it did benefit from the sudden increase in world prices, since all of its sales contracts had clauses allowing revision of prices on a fixed monthly or quarterly basis.[8] For the year 1979 the value of Venezuelan exports climbed to Bs 58 billion an increase of Bs 21 billion, or almost 60 percent over the value of the same volume of exports for 1978. Income per barrel produced went from Bs 49 per barrel to Bs 70 per barrel. This sudden increase in the Venezuelan oil income was clearly a by-product of political upheaval in the Middle East, hence the result of more or less fortuitous events over which the Venezuelan oil industry had no control.

The 1973–74 oil crisis had also been the result of political events in the Middle East.[9] Even the 1970 Libyan events, which had given oil-producing governments formidable new power, had had a dominant political component. Venezuelan politicians could not fail to see the connection between these political events and the increase in oil prices, and some of them started to wonder whether the oil industry really needed, as claimed, highly specialized management. With the country already over the psychological hump of nationalization and now full of self-confidence in its ability to run the oil industry, the pendulum of public opinion slowly started to swing from a position of total support of the industry to a position of critical monitoring. Increasing operational costs and manpower in the industry started to be noticed and talked about. The old feud between some politicians and technocrats took wing. Members of Congress like Ramón Tenorio Sifontes accused Petróleos de Venezuela of "not consulting with Congress some of the changes in their internal organization."[10] Independent petroleum experts like Aníbal Martínez complained about Petróleos de Venezuela "producing more oil than ordered by the Ministry of Energy and Mines."[11] Journalist Federico de Castro declared that the oil price increases "had to do with OPEC decisions and not with an increase in the productivity of the industry."[12]

Still, the predominant mood was one of pride in and of respect toward the oil industry. The new president, Luis Herrera Campíns, celebrated the 1979 increase in the oil prices as a sign of ability on the part of the new administration.[13] The blind faith in the good fortunes of Venezuela as an oil-producing country, the belief that the country would always have a magic source of income and that oil would always be there to bail the country out, seemed to have taken root in the minds and hearts of many of the Venezuelan people.

A Change of Government

In December 1978 the country had national elections for president and congressmen. The campaign leading to this event had been as long as the one resulting in the victory of Carlos Andrés Pérez in 1973, but much less dynamic, a reflection of the main personalities in the race. Both leading candidates, Luis Piñerúa of Acción Democrática and Luis Herrera Campíns of COPEI, were rather colorless politicians, devoid of the strong charisma which Pérez had. The results of the elections confirmed the trend, set since the 1963 elections, that the party in power would lose and the main opposition party would win. The democratic governments since 1958 had been essentially undistinguished and the people had grown accustomed to patiently waiting 5 years to vote against them and replace them with the other main party, hoping that they would have learned their lesson. This always proved to be wishful thinking. The victory of Herrera in 1978 took much of the political world by surprise, which was, in itself, surprising. They apparently failed to see that Venezuelan voters were predominantly independent and that they utilized their voting power to reward or punish the performance of the party in government. The performance of the Leoni, Caldera, and Pérez administrations had been clearly unsatisfactory and, therefore, they had been punished.

In January 1979 the probable choice of Luis Herrera Campíns to become minister of Energy and Mines, Dr. Humberto Calderón Berti, appeared on television and answered many questions on how he would conduct his duties if named to that job. In this interview, Calderón Berti demonstrated an excellent grasp of the situation, giving his answers in technical and precise language. Among other things, he said that "the Venezuelan oil industry would be kept free of political pressures. Recruiting and promotion will be done on a meritocratic basis. . . . The industry should be managed as a private concern."[14] Therefore, when President Herrera Campíns confirmed that Calderón Berti would be his minister of Energy and Mines, most people in the industry sighed with relief. "Here is a man," they said, "who can speak our language."

Humberto Calderón Berti was a 38-year-old geologist with a master's degree in petroleum engineering from the University of Tulsa. After graduating as a geologist in 1964 from the University of Caracas, he joined the Ministry of Energy and Mines. One year later, in January 1966, he was sent to the University of Tulsa. When he returned, degree in hand in 1968, he joined the ministry staff in Caracas as a staff geologist. He resigned less than two years later in April 1970 to become director of the Petroleum Engineering School at the Universidad de Oriente in eastern Venezuela. In January 1973 COPEI's Energy and Mines minister, Hugo Pérez La Salvia, named him director of reversion, a new department in the Ministry of Energy and Mines. In May 1976 he became executive vice-president of INTEVEP.

This was the professional experience of the man who had become the minister of Energy and Mines in the cabinet of Luis Herrera. Calderón Berti had also had many years of political activity since his high-school days. In the course of those years he had become a close personal friend of the new president, Luis Herrera. He was dynamic, creative, and had shown an excellent grasp of the main policy issues of the oil industry. He had a sociable attitude and rapidly won the acceptance of the oil-industry community.

In early 1979 oil-industry employees organized in AGROPET sent President Herrera and Minister Calderón Berti a document containing some opinions on the most pressing issues that the new government would have to face in the field of petroleum. It might be worthwhile to transcribe some of the concepts advanced in that document.

On Petróleos de Venezuela:

We believe that the balance of the first four years of existence of this holding company has been positive . . . especially useful has been its role as a cushion between the technical sectors of industry and the political groups.

On politicization:

It will be vital to demonstrate your support for the principle, already contained in your program, that the industry will remain free from politics and from the practices of cronyism which have proved so harmful to other sectors of public administration.

On the new board of PDVSA:

> We recommend making up the new board of the holding company with a greater share of managers promoted from within the industry and to limit the number of directors coming from the outside.[15]

It is evident from these excerpts that the issue of political interference in the management of the national oil industry still was one of the main concerns of the oil-industry employees in 1979. From 1976 to 1979, success seemed to have been due to the loyalty shown both by oil managers and by the political sector to five basic concepts: self-financing, freedom from political interference, career development based on merit, professional management, continuous normal operation. As long as the industry adhered to these principles, it would stay healthy and profitable.

The Shareholder's Meeting of PDVSA, August 1979

The New Board. As the time to renew the board of PDVSA drew near, it became clear that the new administration had not made a decision on the subject. Suspense mounted within the oil-industry ranks because selection of key personnel had always been an orderly and predictable event. Replacements were usually known much before they actually took place. Organizations determined promotions according to a strict set of procedures based on personnel ranking and evaluations. It was said of the Shell organization that "whenever a managing director retired, a new office boy was hired," since everyone in the organization would receive a promotion. This jest illustrates the type of careful, conservative management style that the Venezuelan oil industry had inherited.

The new minister, H. Calderón Berti, broke with this practice. Up to a few hours before the new members of the board were to be named, no one knew who they would be. Either the new administration preferred not to disclose the names beforehand, or the decision had not yet been made. Oil-industry personnel felt that they had reasons to worry, whatever the true explanation.

One immediate consequence of the mystery surrounding the naming of the new board was a significant increase of tension within the small group of top oil executives, especially those at PDVSA. Another was the appearance of self-nominated candidates for the would-be vacant jobs. Among these was Eduardo Acosta Hermoso, an engineer who had been head of the Petrochemical State Company in the 1960s. Dr. Acosta Hermoso had been named by President Herrera commissioner for environmental protection, with particular emphasis on oil-industry operations. This position was largely symbolic but Acosta Hermoso used it to travel extensively throughout the oil fields, meeting with managers of the industry and suggesting that he could become the future president of PDVSA.[16] The unusual campaign of Dr. Acosta Hermoso was not taken seriously at high levels but helped to create considerable confusion among the younger managers and engineers of the industry.

General Alfonzo Ravard wished to continue as president of PDVSA. He also went campaigning. Usually a very reserved, aloof man, he became extremely active socially. He was seen at all events of the industry. He mobilized his considerable supporters in Fedecámaras, in government, the armed forces, and other important sectors, to pave the way for his reelection. Since he had done an excellent job during his first tenure and, under him, the industry had achieved extremely good results, his reelection did not seem illogical or undeserved.

There were, however, people who firmly opposed the General's reelection. Some of these people were at high executive levels in the operating companies. They thought that the basic mission of Alfonzo Ravard had already been accomplished. He had been put in charge of PDVSA during a first stage of consolidation, which had worked out extremely well. His presence had served to eliminate the possible negative political impact of placing at the helm of the nationalized oil industry some of the Venezuelan executives who had been associated for years with the multinational oil companies. But now, 4 years later, this risk had disappeared and the oil-industry community had strong expectations of seeing one of their leaders get the top job. For them the continued presence of Alfonzo Ravard in the presidency of PDVSA seemed a step in the wrong direction. They were pragmatic enough to realize however that if the job was going to be given to an outsider, then Alfonzo Ravard was still the best choice.

Another influential person who seemed to favor the replacement of Alfonzo Ravard was the new minister, Calderón Berti. This might sound surprising, but it was easily explainable to those who knew the new minister well. He was a relatively young man, suddenly thrown into a position of considerable power. He felt lonely and not very secure. During the previous 2 years, he had been part of the PDVSA's system at INTEVEP, where he had been treated by Alfonzo Ravard with aloofness when not with distrust. Now it was his opportunity to reply in kind and at the same time place at the helm of PDVSA someone to whom he could relate with more ease and rapport.

The two front candidates from industry were Guillermo Rodriguez Eraso from Lagoven and Alberto Quirós from Maraven. Most people took it for granted that Rodriguez Eraso would be chosen over Quirós because of seniority considerations and also because of his more conservative, cautious management style. Quirós was more creative, perhaps the best strategist in the industry. He also cut a high profile which was not to everybody's liking. His brash manners in PDVSA's top-level meetings had made him both a respected and resented figure.

The decision was finally made to keep General Alfonzo Ravard as president of PDVSA for another period. He had proved again to be a superb survivor, prevailing over strong opposition. Likewise vice-president Julio César Arreaza was retained. Arreaza had done a reasonably good job. During the first four years he had become the man who ran the day-to-day administrative side of the company. He was careful, paid considerable attention to details, and was much better organized than General Alfonzo. Keeping Arreaza, however, introduced a change in the unwritten rules of the game. Arreaza had originally been chosen as vice-president because of his dual roles as administrator and politician, closely identified with Acción Democrática. The government thus indicated that it would consider the position of vice-president of PDVSA as the liaison between the industry and the political world. But in order to do this effectively, the vice-president had to be closely identified with the party in power, had to be able to communicate with key political leaders in that party and have their confidence and trust. After the defeat of Acción Democrática at the polls in December 1978, the observers of the oil industry became convinced that Arreaza would be replaced in August 1979 by a COPEI sympathizer. Rumors were strong that Antonio Casas Gonzalez, a former minister of planning during the administration of Rafael Caldera, and with a long association with the Interamerican Development Bank, would be Arreaza's replacement. Casas Gonzalez, in fact, became a member of the new board, but Arreaza stayed as the vice-president.

There were six new members of the board. Four were from outside the industry: Wolf Petzall, Humberto Peñaloza, Antonio Casas Gonzalez, and alternate director Manuel Pulido. Two were promoted from within the industry: Gustavo Gabaldón and Hugo Finol. The repeating members of the previous board were Rafaél Alfonzo Ravard, Julio César Arreaza, Pablo Reimpell, Alirio Parra, Manuel Ramos, Edgar Leal, and alternate director Luis Plaz Bruzuál. Although the minister had had a deciding influence over all of the new nominations, he seemed to have been especially responsible for two of them: Petzall's and Peñaloza's. Petzall was a geologist and had taught Calderón Berti geology at the University of Caracas. He had worked for Creole Petroleum (Exxon) for several years in the planning area and, at the time of his resignation, in the public relations field. He went to Corimon, a private group in Venezuela active in petrochemicals, paints, and foodstuffs, as an executive assistant to the president. He was at that position when Calderón Berti asked him to join PDVSA.

Peñaloza was a petroleum engineer who had had an early and brief career in the oil industry with Creole. He had gone, also briefly, to the national telephone company. Eventually Peñaloza had formed a small company called Mito Juan, which produced oil from marginal oil fields and which also had a drilling division. Peñaloza was an entrepreneur with great concern for community affairs. He had been a promoter of musical groups and had founded a nonprofit, cultural FM radio station, the only one of its kind in Venezuela. Peñaloza was also very politically inclined and it was evident that his only moderate success in business and politics was due to the fact that he had never been able to make a clear choice between these two fields. Up to this moment, Peñaloza had been a strong critic of the industry and the minister perhaps hoped, by naming him, that he would be able to put into practice the improvements he had recommended from the outside.

Antonio Casas Gonzalez was a public manager of experience, well-known in the economic and planning fields. He was closely identified with COPEI, so there was no doubt he would play a role with COPEI similar to the one played by Arreaza vis-à-vis Acción Democrática. Manuel Pulido was a young engineer who had been active in the petrochemical and aluminum fields. Several years earlier, he had been the general manager of the ill-planned, aborted joint venture between the Ministry of Energy and Mines and British Petroleum for the construction of a bioprotein plant in Venezuela. Pulido's presence on the board perhaps had to do with his petrochemical experience but was probably partly due to his political connections, his sister being a member of the cabinet.

Gustavo Gabaldón came from Maraven, where he had been corporate director responsible for personnel matters. Gabaldón was a lawyer from the Trujillo State in the Venezuelan Andes, the same region as Calderón Berti, and had considerable experience in the field of labor relations. He was a dynamic manager, very direct in his approach, independent, and very ambitious.

Hugo Finol also came from Maraven, where he had been corporate director responsible for the areas of exploration and production and later trading and supply. Finol was a mechanical engineer with a quiet nature and a very calm approach to his business. He was highly regarded in the industry for his experience in production.

The continued career of Pablo Reimpell should be mentioned. He had become perhaps the most useful and versatile member of the board of PDVSA. A very solid financial expert, he was also conversant with corporate planning and had an almost infinite patience, which helped him to get things done in spite of the frequently unsettled atmosphere of PDVSA.

Changes in the Bylaws of Petróleos de Venezuela. The shareholders' meeting of August 1979 was especially significant because executive decree 1123, dated August 30, 1975, containing the bylaws of Petróleos de Venezuela, was partially modified, as follows:

Shareholders would now have the authority to assign the areas of managerial responsibilities to all members of the board.

The board would have two vice-presidents instead of one.

The members of the board would have a tenure of 2 years instead of 4.

Shareholders would have the right to approve or disapprove the consolidated budgets of the holding company and those of the subsidiaries. To this purpose the budgets should be sent to the Ministry of Energy and Mines at least 30 days before the date of the assembly.

In the Caracas English-language newspaper, journalist Everett Bauman stated his opinion:

> the modification which requires the approval of the budgets of the subsidiaries of Petróleos de Venezuela by the shareholders (of Petróleos de Venezuela) is an unfortunate one. PDVSA generates their own income and their operational decisions must be made on a strictly technical basis. If the government is not pleased with the administration (of Petróleos de Venezuela), they can change it. . . . The approval of the yearly budgets at this level can open the door to political interference.[18]

An analysis of the changes, made by the magazine *Resumen,* agreed with the opinions set forth by the *Daily Journal* and added,

> Although we believe that such a tool in the hands of the new minister of Energy and Mines will be wisely utilized, there is the danger that such procedures would, in the long term, reduce the efficiency of the industry. . . . The specter of the petrochemical and CVP disasters is still fresh in our memories.

Resumen said further,

> the modification by which the shareholders' assembly can now assign managerial responsibilities to the members of the board goes much farther than it would seem desirable from a healthy organizational standpoint, because these are decisions best left in the hands of the corporations themselves.

Resumen concluded its commentary by saying, "Let us wait and trust."[19]

The industry received these modifications with mixed feelings. The executives and managers of the operating companies definitely did not like the trend suggested by the modifications. At the same time the initial impression made by the new minister on the industry had been so favorable that most employees were prepared to give him the benefit of the doubt. The oil industry adopted a wait-and-see attitude, reinforced by the deeply ingrained discipline typical of these organizations for so many years.

To appreciate the significance of these changes, it must be remembered that the shareholders' representative was the minister of Energy and Mines. His decisions became the decisions of the nation in the field of hydrocarbons. Rarely had a minister of Energy and Mines in Venezuela had so much power, and rarely had power been in the hands of someone so young and inexperienced. Still, Calderón Berti was not perceived as a risk by industry management because he appeared to listen and seemed to invite participation. He now had authority to tell PDVSA board members what tasks they should be responsible for, to remove them or name new members every 2 years, and he could now look at the capital and operational budgets to suggest or order changes.

There is no question that these changes in the work mechanism of PDVSA introduced a substantial departure from at least two of the five basic strategic concepts on which the objective of a healthy nationalization had been based, namely, freedom from political interference and professional management practices.

In a similar manner the reelection of General Alfonzo Ravard as president of PDVSA was seen by many oil-industry employees as an unwelcome departure from the stated original intentions of giving that job to a top executive promoted from their ranks. The selection of some of the new directors did not seem to follow proper evaluation procedures but seemed to have been made on the basis of Calderón Berti's personal preferences. This could eventually lead to an undesirable situation, the emergence of two groups within the board: a group loyal to the industry and one loyal to the minister. Naming new directors every 2 years could lead to fighting for power and to increasing disharmony within the small group of hopefuls. Having the budgets of the operating companies approved at the political level invited inefficiency and politicization because the ministry simply did not have a group of experienced analysts to evaluate these budgets.

The Industry under Fire, 1979–1980

Although the oil industry continued generating increasing volumes of money for the government, its operating costs increased, basically due to inflation and the higher level of activities. Because of the increasing costs, the political sector became more concerned about the degree of control government should have over PDVSA. Under the presidency of Carlos Andrés Pérez, the oil industry had been given almost total autonomy. Oil-industry management had found in Minister Valentín Hernández a man not given plays or to exercises of authority. He wanted above all to keep the industry working smoothly and free from conflict.

Calderón Berti did not seem to think in the same terms. He seemed determined to have the power he felt Hernandez had not wanted. His performance during the shareholders' assembly of August 1979 clearly suggested a trend of rapidly tightening controls of the industry on the part of the Ministry of Energy and Mines. In this effort, Calderón seemed to be rather enthusiastically supported by a small core of ministry officials who felt very strongly that power should have never shifted from the ministry to PDVSA. They wanted the pendulum to swing back.

The desire for increasing control of the oil industry did not seem restricted to Calderón Berti. It was also very much in the minds of the political parties. The parties vigorously tried to increase their monitoring of oil-industry activities and decisions through congressional committees and task forces. The major complaint of the legislators about the oil industry was what they termed its extreme degree of secrecy. The political sector started feeling left out and insufficiently informed of what the oil industry was doing. While the industry kept showing increasing profits and the country was still undergoing the psychological effects of nationalization, the results seemed to suggest that oil management decisions were probably more responsible for the negative aspects of the industry (increasing costs and manpower) than for its positive aspects (increasing income). Much of the glamour and prestige with which top oil management had been perceived by many political leaders started to wear off. Complaining about "insufficient" information given by industry to Congress, Acción Democrática petroleum spokesman Celestino Armas said: "We are to blame because we (originally) gave them excessive freedom."[20] Perhaps no project of the industry illustrates more clearly the increasing degree of friction and distrust which emerged between the oil industry on the one side and the government bureaucracy and the political world on the other, than the Orinoco heavy-oil development project.

The Orinoco Heavy-Oil Development Project. The immense volumes of heavy oil in the subsurface of the Orinoco area of Southeastern Venezuela had long been recognized by the industry as a major source of future hydrocarbons for a country which had already seen its traditional oil fields lose considerable vigor and production capacity. These deposits had been located in the 1930s. It was not until the late 1960s and 1970s, when the increasing prices of oil in the world markets made them look economically attractive, that renewed efforts were made toward their evaluation and development. The task of evaluating the Orinoco heavy-oil deposits had originally been given to a unit created at the Ministry of Energy and Mines, small in size and subject to the chronic problems of shortage of funds and human resources typical of government agencies. This unit started its evaluation work in the early 1970s but had not made much progress by the time of nationalization.

With the creation of Petróleos de Venezuela, the Venezuelan oil industry acquired a general coordinating and planning organism which could look at the needs of the industry in an integral manner. This required that the Orinoco heavy-oil development project be transferred from the ministry to PDVSA. However, there was fierce opposition from the ministry to accepting this transfer. What seemed to make sense from an organizational viewpoint did not make sense, at least to the ministry, from the viewpoint of control. To give up control of the Orinoco project was, for the ministry, to lose considerable power. The ministry officials who wished to retain this control enlisted at least one important group to lobby in their favor: the National Energy Council, an advisory body of the president for energy matters, made up of representatives of the political parties and of economic groups such as Fedecámaras and Pro-Venezuela. This organism became very vocal in denouncing what they termed the ravaging of the ministry at the hands of PDVSA. At the heart of their position there seemed to be some deeply rooted ideological motivations. The Energy Council was essentially dominated by the political left and by ultranationalistic members. They perceived the ministry staff as the only true defenders of the hydrocarbon resources of the country. To them, handing over the Orinoco project to PDVSA was practically an act of treason.

Typical of this way of thinking were the opinions of Radamés Larrazabal, member of the Venezuelan Communist party and of the National Energy Council. Larrazabal suggested that assigning the Orinoco project to PDVSA and to the operating companies was "identical to giving it to Exxon, Gulf, Shell, and Mobil." Larrazabal had also written a previous letter to Minister Valentín Hernández warning him that giving the Orinoco project to PDVSA would mean "giving control of the project to the multinational oil companies." He suggested that "the state should completely resume all exploration work, utilizing the knowledge of Venezuelans or technologies obtained in state-to-state negotiations."[21]

Although the Orinoco project was finally handed over to PDVSA through the decisiveness of Minister Valentín Hernández, the open fight that had been put up by ministry officials and their political allies marked another step in the politicization of the industry. The Orinoco area was so large that no one operating company by itself could handle it properly. It was therefore decided to divide the area in four portions, each to be explored and evaluated by one of the four operating companies. The eastern-most portion, called Cerro Negro, was assigned to Lagoven. This area was not really unknown, being located very close to existing producing oil fields and having been subject to some preliminary drilling, in 1970–71, which had confirmed the presence of heavy-oil deposits south of the Morichal oil field. This area was therefore the logical choice for a production project.

Lagoven believed that the thicknesses of oil sands in the Cerro Negro area and the presumed extension of the oil deposits could easily sustain a production of some 180,000 barrels per day and probably could increase to 500,000 barrels per day. The original idea was that the 180,000 barrels per day which could be produced from this area would be upgraded to produce some 125,000 barrels per day of a commercially acceptable crude oil by 1988. These plans of Lagoven were approved both by the Ministry of Energy and Mines and by Petróleos de Venezuela as part of the long-range planning of the industry. The project would be both expensive and complex. It included considerable drilling, the erection of upgrading plants, electrical generating facilities, new pipelines, and a possible oil terminal on the Orinoco river.

Guillermo Rodriguez Eraso, president of Lagoven, explained to Jorge Olavarría, editor of *Resumen,* how planning for the project was done. Once the terms of reference were obtained from the Ministry of Energy and Mines, he said,

> We engaged in the planning. We formed a group of our own people . . . fifteen or twenty . . . to do the preplanning. . . . Once we reached an understanding of what to study in detail . . . we thought we should ask for help because the task was too big for us. We contacted a number of qualified companies . . . made a careful analysis of all the offers we received (from twelve companies) and finally chose one, Bechtel, to become our advisor in mid-1979.[22]

He defined the relationship of Bechtel with Lagoven in the following manner:

> We formed a committee managed by our own people and task forces for each function . . . for example, refining. These task forces could be led by a member of Bechtel or could be led by a member of Lagoven . . . but management (was) totally in our hands. The planning group had two main objectives: a master plan for the execution of the work and a budgetary estimate.

Rodriguez Eraso added that the planning studies were to be finished by the end of 1981 or early 1982 and that they would involve some 1.2 million man-hours of work. The costs of the planning effort alone were estimated by the president of Lagoven in some Bs 350 million ($80 million). Lagoven added Rodriguez Eraso, "had also engaged a large number of Venezuelan consultants and advisors, including several universities, to help in the planning stage."

By mid-1980 it became evident that if Lagoven wanted to complete the project by 1988, it had to start on the construction stage without waiting for the planning stage to be fully completed. There had to be an overlap. Rodriguez Eraso said,

We opened an international bidding process (to choose the general coordinator of the construction stage) inviting eighteen companies from the United States, England, France, Italy and Japan. . . . The recommendations of the evaluating group went to the project management group . . . from there, up to Lagoven's central contracting committee headed by the finance manager of the company . . . the recommendations were analyzed at this level, confirmed, and brought to the attention of Lagoven's board. . . . Once approved at board level, the recommendations went to the holding company. . . . There, too, at board level, the decision of Lagoven was backed. . . . Finally, it went to the National Contracting Committee of the Oil Industry, a group headed by the legal advisor of Petróleos de Venezuela and made up of several members of the board of Petróleos de Venezuela and all the presidents of the affiliates of PDVSA. . . .

This is the way Lummus was selected. Of eighteen companies, only five remained at the end: Lummus, Bechtel, Fluor, Davy McKee, and Technip. The two latter companies had no chance and were notified accordingly . . . the other three were in the contest up to the last moment, when the final recommendation was made.[23]

The decision favoring Lummus was based, according to Lagoven, on the following points: Lummus offered the best combination of economic terms and capacity to coordinate the execution of the project. Lummus offered the most coherent, qualified, and solid managerial team. And the better integration of the associated Venezuelan company, Vepica, in the overall organization would allow for an increased participation of the Venezuelan company in the execution of the project.[24] Lagoven estimated that the fees to be paid to these companies would amount to Bs 3 billion over a period of about 9 years.

While this process of evaluation for the most important project of the Orinoco area was going on within Lagoven, political opposition to the control of the Orinoco heavy-oil deposits by PDVSA was increasing. Already in March 1979, Aníbal Martínez, a COPEI sympathizer, oil expert, and advisor to Pro-Venezuela was defining the work to be done in the Orinoco area as "useless" and lamenting "the loss of authority of the technical staff of the Ministry (of Energy and Mines) vis-à-vis oil-industry employees who should be supervised and controlled."[25] Martínez was also quoted as saying that "each of the (operating) companies is trying to convince Petróleos de Venezuela to give it money to get a head start for the Orinoco area."[26]

The political parties, including COPEI, started to be very critical of the way Lagoven was handling the Orinoco Cerro Negro project. Their major concern was that Bechtel, the company advising Lagoven in the planning stage, was perceived as a politically flavored multinational, as dangerous as Exxon or Shell. Prominent political figures in the United States such as George Bush, Caspar Weinberger, and George Schultz had been associated with Bechtel. The role played by this company in the development of the

Orinoco area was seen as a ploy of the U.S. government to obtain control of the huge heavy-oil deposits present in this area. Lagoven management, in turn, was seen almost as an accomplice to this confabulation. Again the major exponents of this theory were the leaders of MAS, who repeatedly denounced in Congress what they termed the surrender of the Orinoco hydrocarbon resources into the hands of the U.S. government. The major political parties, although not as vociferous as MAS, also joined in denouncing what they saw as U.S. government pressure for an accelerated development of the Orinoco heavy-oil deposits, a step which in their view was not desirable for Venezuela.

In fact Venezuela clearly seemed to have no other choice but to develop the Orinoco heavy-oil deposits if the country wished to avoid serious balance of payments deficits by the 1990s. More than to the United States, the development of the Orinoco oil was vital to Venezuela.

The opposition was also fueled by the belief that Bechtel would also go on to obtain the contract for the construction phase and that its fees would amount to Bs 20 million or even 30 million. Guillermo Rodriguez Eraso, the president of Lagoven, answered these charges by saying:

> The truth is totally different . . . the big project is made up of a series of (smaller) projects. The winner (of the contract for the coordinating role in the construction phase) is going to be a coordinator but is not going to actually engage in construction. . . . Construction will be done by other companies, hopefully Venezuelans in their great majority.[27]

The perception of many political sectors and even of the ministry technical staff that Bechtel had a decisive advantage in the race to win the contract as general coordinator for the construction stage of the Cerro Negro project was not illogical. Bechtel had been cooperating with Lagoven in the planning stage, and was therefore fully conversant with the project—knew it inside out—and had established close contacts with Lagoven's top management. Ordinarily the role played by Bechtel in this stage should have prevented this company from participating in the bid for the construction stage. However, Bechtel had requested to be allowed to participate in the construction stage as a prerequisite to playing an advisory role in the planning stage and Lagoven management had agreed to this. Several members of the National Contracting Committee of the oil industry had voiced their deep concern about this situation, which they believed could place Lagoven and the industry in a vulnerable situation vis-à-vis public opinion. At the time, however, the predominant opinion in PDVSA and Lagoven was that this should not pose a major problem. It is apparent that they underestimated the critical approach that the political and public opinion were taking toward the handling of this project. They also underestimated the strength of the perception that Bechtel was an arm of the U.S. Department of State.

The ministry staff started to pressure Minister Calderón Berti to order Lagoven to modify the Cerro Negro project designed to give Venezuelan engineering companies a larger role to play in the control of the project and give back to the ministry the authority it had lost. Within the ministry, however, there were very serious signs of administrative deterioration. During 1980 Calderón Berti had dedicated almost all of his time to international oil matters and had neglected the strengthening of the professional staff at the ministry. A group of ministry employees denounced this situation publicly:

> Every day it becomes more evident—they claimed—the lack of a coherent administrative activity (in the ministry) and this brings about a deterioration of the role that the Ministry of Energy and Mines should play. . . . there is no such thing as an integral mining and energy policy. . . . The activity of the ministry is oriented toward the issues of international petroleum policy, neglecting the internal energy situation.

The ministry employees accoused Calderón Berti of not working with his subordinates:

> At the top level of the ministry there is lack of communication with the technical levels and this generates a climate of uncertainty and explains the absence of plans and work programs.

The employees also mentioned "the lack of austerity" at the high levels of the ministry and stated that because of all of these circumstances, "more than fifty specialists had resigned in a 3-month period."[28]

It is obvious that an organism in crisis such as was the case with the Ministry of Energy and Mines could not hope to take control of a huge project like the Orinoco heavy-oil development project. However, Minister Calderón Berti initiated at ministry headquarters a series of what he termed "consultation meetings" about the Orinoco project with political parties, engineering and geologic societies, and other public groups. This opening gave those sectors opposed to the Orinoco project a new platform to voice their disagreement. In these meetings, especially those held with the Venezuelan Engineering Society and with political parties such as MAS, the management of Lagoven was subject to very harsh criticism for their perceived role in favoring Bechtel. Sectors of public opinion started to claim that the Orinoco projects should not continue unless there was general agreement among the public about their desirability. The danger of this situation was obvious. The projects of the oil industry could not be subject to the approval of political parties, professional societies, and pressure groups without a dramatic loss of efficiency. Even a suggestion that this should be so amounted to interference in the management of the oil industry. The lack of judgment shown by Lagoven's management in accepting Bechtel's request to participate in the bidding of the construction stage of the Cerro Negro project and the political opportunism shown by Calderón Berti in trying to capitalize this error to gain prestige and popularity among the political parties and other public groups converted what should have been a very professional process to a political three-ring circus.

Even after Lagoven made public its decision to select Lummus for the general coordinating role of the construction stage of the Orinoco project, the political parties and Calderón Berti were certain that, had not it been by their intervention, Bechtel would have been selected and the U.S. Department of State would today be in control of the Orinoco heavy-oil deposits.

Other Signs of Disharmony. Although the open fracas over the Lagoven Orinoco project was the most notorious, the oil industry was experiencing very intense pressures and criticism from many outside sectors. In March 1979 Hugo Pérez La Salvia, minister of Energy and Mines during the administration of Rafael Caldera gave an extensive interview to a Caracas magazine, in which he made some highly critical comments regarding the performance of the nationalized oil industry. Pérez La Salvia said, among other things:

> I have always said that with the advent of the so-called nationalization we inherited the management of the multinationals and I think that these managers already had a mentality derived from their work with the concessionaires. This situation must change.

Pérez La Salvia added, in connection with the salaries of oil industry managers:

> The salaries of the managers of the nationalized operating companies are too high. . . . It is unfair that the employees of the Ministry of Energy and Mines, who have in their hands the control of the industry . . . receive salaries several times lower than those of the managers of the operating companies.[29]

Everett Bauman, writing in the same issue of the magazine about Pérez La Salvia's comments, said,

> It seems that Pérez La Salvia . . . with the support of some of his COPEI party colleagues, is advancing the thesis that the oil industry should be under political and government control. . . . The country depends largely on the oil industry and will continue to do so for many more years. Venezuelans are conscious of this fact and wait anxiously the decisions of . . .

President Herrera Campíns about the manner in which he will deal with the industry to keep it efficient and free from political interference.

COPEI sympathizer and oil expert Aníbal Martínez, already cited in this chapter in connection with the Orinoco project, also had things to say about other aspects of the industry. In a summary of the findings of president-elect Herrera's oil study group, of which he was a member, Martínez wrote as follows:

> The problems of the Venezuelan oil industry are of a different nature and all are due to one single reason: the decrease in hierarchy (power) of the Ministry of Energy and Mines.

He added,

> This is the source of the delay in the delivery of the policy guidelines in hydrocarbons from the Ministry to Petróleos de Venezuela, of the overproduction (of oil) as compared to the real capacity of the reservoirs . . . of . . . the individual programs, subject to *forced* approval, to modify the refineries' product mix, of the useless partition of the Orinoco heavy-oil belt in four segments . . . of the opening of four simultaneous fronts for exploration in the continental shelf, perhaps in unfavorable geological locations . . . of the unrealistic analysis on how to proceed in the local market, of the lack of interest in the development and recruiting of adequate human resources, of the strange competition between sister companies, of the loss of authority of the technical staff of the Ministry (of Energy and Mines) vis-à-vis the employees of the oil industry . . . of the obstacles placed in the way of locally generated oil research, of the continuous implementation of the technological agreements which have been generally rejected.[30]

Coming from a man familiar with the industry, these charges were very serious ones and were very representative of the type of opposition the industry was increasingly getting.[31] Essentially, the arguments were that the oil industry was not being properly controlled and that many of its activities did not seem to be in tune with national objectives. How valid were these arguments?

Loss of Control of the Ministry of Energy and Mines over the Industry. This argument was only partly valid. If loss of control meant a decrease in the technical audit capacity of the ministry over the industry, this was not true. The ministry had all authority in this field and exercised it fully. Ministry staff ordered oil production open or shut in, approved the technical aspects of production and refining projects, and generally made certain that the operating companies abided by the regulations of the hydrocarbons law. The only limitation to the exercise of this authority was in the quality of the ministry's staff. In this field they did have problems. The staff of the ministry had been getting weaker since nationalization. Technical staff preferred working in industry, where salaries and general working conditions were much better. As a result the staff in the ministry was not of the first caliber. Several of those who had remained seemed to be motivated by feelings of resentment against the companies, while others were fueled by deeply nationalistic feelings. All tended to have a systematic distrust of the industry.

Now, if loss of control meant decreased managerial control of the industry, then the argument was true. The ministry did not have managerial control of the industry. It was not supposed to have it. The role to be played by the ministry in the nationalized oil industry was not a managerial one since this role was clearly reserved to PDVSA and the operating companies. Decision-making for the oil industry was never in the hands of the ministry, except in a partial way, in the years between 1972 and 1975, when control of the industry was being taken out of the hands of the concessionaires and Petróleos de Venezuela had not yet been formed. Once Petróleos de Venezuela was formed, there was no logic in the continued desire of the ministry staff to keep control of the decision-making mechanisms.

What was perceived as loss of control of the government sector over the industry was in many ways a consequence of the increasing internal deterioration taking place at the Ministry of Energy and Mines. This had resulted in a decrease of the analytical power of the ministry staff to process efficiently the information received from the industry.

Confusion in the Policy Guidelines Given by the Ministry to Petróleos de Venezuela. This argument was not valid. Most people outside the industry believed that the policy guidelines in the hydrocarbons field were drawn up by the ministry staff. This was not the case. There was no one at the ministry doing this work, which required efficient planning and careful staff work. Apart from very broad guidelines provided by the ministry's top officers (regarding level of production, for example), detailed policy guidelines were drafted in the industry and then sent to the ministry. Once the minister had seen and approved them, they would be sent back to the industry as the official policy guidelines of the Venezuelan government for the hydrocarbons sector. The guidelines represented the generally agreed objectives to be pursued by the industry and were totally consistent with national objectives. Otherwise they would have never received the approval of the minister. There was no truth, therefore, in the claim that Petróleos de Venezuela did not pay attention to the guidelines. It would have certainly been most illogical not to agree with one's own guidelines.

Also in March 1979 Humberto Peñaloza, who later would become a member of the board of PDVSA, defined the technical support agreements signed with the multinational oil companies as "obstacles in the path of (a true) nationalization." He claimed that "via the technical support agreements, the multinational companies participate in the planning of the industry at the highest levels." He said that the existence of these agreements in each of the four operating companies "helped to explain the partition of the Orinoco area into four segments." Peñaloza added:

> An external observer has to admit there is evidence of important shortcomings in the planning, coordination, and control in Petróleos de Venezuela. A careful analysis of what is taking place points to only two explanations: either the interests of the operating companies are prevailing over the general interests of the industry and the country, or things are taking place without the knowledge of the holding company.[32]

These arguments were very similar to those expressed by Aníbal Martínez. Both of these critics of the industry were advisors to Pro-Venezuela and leading members of a small group calling themselves "Front for the Defense of Petroleum." At the core of their arguments was the perception that the oil industry had broken free of control and was now acting in a manner not consistent with the national interest. This perception was strongly reinforced by the distrust they felt of top oil industry managers. Peñaloza and Martínez, together with other Venezuelans such as Pedro Márquez, Eduardo Acosta Hermoso, and Alfredo Tarre Murzi had been part of the Pérez Alfonzo team in the early years of OPEC. They felt very strongly that they had been the true nationalists while other Venezuelans had preferred to keep working for the multinational oil companies. When nationalization of the Venezuelan oil industry took place and the men who had worked for years with the multinational oil companies had been retained as the top managers of the industry, men like Peñaloza and Martínez probably felt that the rewards had gone into the wrong hands. This probably led them to the adoption of a very critical attitude toward the decisions being made by the managers of the industry.[33]

Other sectors kept claiming that Congress should have much more control of the budget of PDVSA. Armando Sánchez Bueno, former minister of justice during the presidency of Carlos Andrés Pérez and at the time vice-president of the National Congress, said publicly that the control of the nationalized oil industry should be in the hands of Congress:

> The Mexican Congress approves the budget of PEMEX, the Mexican state oil company, while here, in Venezuela, a country with considerable legislative experience, we have not been able to obtain that state-owned enterprises submit their investment proposals and budgets to Congress for approval. We should pass a law in this respect.[34]

The New Environment

In the period 1979–80, several factors seemed to combine to cause significant deterioration of the relationship between the nationalized oil industry and the political sector of the country:

The change in government which took place in March 1979

The modifications made to the bylaws of Petróleos de Venezuela and the change in management style which seemed to accompany those modifications

The acute economic problems of the country, which made the political sectors much more conscious of the shortcomings of the industry as well as of the existence of an $8 billion oil fund which could be utilized for other purposes[35]

The continued criticism coming from nationalistic groups which did not trust top oil management and from ministry staff desiring to regain the power they lost at the time of nationalization

The increasing political posture of the new minister of Energy and Mines, Humberto Calderón Berti.

These factors combined with technical problems to make life much more difficult for the oil industry technocracy. They now faced immense challenges such as the need to manage efficiently the large projects which were underway and the need to increase production potential to 2.8 million barrels per day while also having to worry about increasing costs and the increasingly hostile political environment.

To put it briefly, the nationalized oil industry was finally being assimilated into the system. For 4 years after nationalization, mostly as a product of inertia, a clear separation remained between the largely disorganized and politicized public sector and the orderly and professional oil industry. During those 4 years the political sectors had been very careful to let the professional managers of the oil industry take their decisions. Now, "the honeymoon had ended."[36]

Most observers claimed that such an outcome was inevitable. It was impossible, they said, that an industry which played such a decisive role in the economic life of the country could have permanently stayed outside the sphere of intense political influence and maneuvering. However, this was the very characteristic of the nationalized oil industry that all political, economic, and professional sectors of Venezuela had solemnly promised to respect when nationalization was decided and implemented in 1975. This had been, in fact, the basic premise for a successful nationalization.

What made these groups forget their promises, change their minds, break their self-imposed rules of the game? Much of the explanation should probably be sought in the tendency of politicians to look almost exclusively at the short term. As Venezuelan political sectors became aware that the Venezuelan oil industry was operating normally, they started to feel that it no longer required a privileged treatment and that in fact the time had come for them to reap part of the profits. In order to do this, however, they had to obtain control of the industry.

What began to take place, therefore, was a power play, an increasingly open struggle for control of the oil industry. The Congress and the Ministry of Energy and Mines became involved in this struggle and the weapons utilized escalated. Many outside groups and individuals were brought into the fray: engineering societies, conservationist groups, regional business groups, even religious leaders. They all claimed to defend the rights of the common citizen against the all-powerful and insensitive oil industry. The technocracy of the oil industry found itself not only faced with the responsibility of generating the income required by the government to sustain national economic growth, but also surrounded by increasing hostility from important portions of public opinion which seemed to have become the tools of political warfare.

Notes

1. R. Alfonzo Ravard "Cinco años de Normalidad Operative," Petróleos de Venezuela, Caracas, 1980.

2. The guests were usually university professors, students, newspapermen, and intellectuals who admired him and took his words as gospel.

3. See *Resumen* 259, for an analysis of the Pérez Alfonzo document. The complete document was never published, although mimeographed copies still exist in many hands, including my own.

4. Dr. J. Pérez Alfonzo died in Washington, D.C. in September, 1979. He left instructions in his will to be cremated and his ashes dispersed over the sea. He did not want to be buried in Venezuela with wasteful and pompous ceremonies.

5. *El Nacional,* May 27, 1979. Reviewed in *Resumen* 293, June 17, 1979, p. 30.

6. *Resumen* 167, January 1977.

7. See G. Coronel, "Oil: Towards the Age of Balance," Annual Report, University of Tulsa, 1982.

8. The prices of some refined products were revised even more often.

9. Coronel, "Oil, Towards the Age of Balance," p. 4.

10. See *Resumen* 305, September 1979, p. 50.

11. Ibid.

12. Reported in *Resumen* 306, September 1979.

13. *El Nacional,* September 30, 1979, p. D-1.

14. "Lo de Hoy," Friday, January 31, 1979, videocassette of this program.

15. "La Industria Petrolera Venezolana en 1979. . . ," a document from AGROPET to President Elect, Dr. Luis Herrera Campíns, Caracas, January 1979.

16. A report in this connection appeared in *Ultimas Noticias,* March 29, 1979, p. 2.

17. Gaceta Oficial de la Republica de Venezuela 31810, August 30, 1979.

18. *Daily Journal,* August 30, 1979.

19. *Resumen,* 306, September 16, 1979.

20. *El Diario de Caracas,* p. 9, April 27, 1981.

21. *El Nacional,* June 8, 1979, p. D-17. The letter was quoted in the newspaper report.

22. *Resumen* 391, April 1981, pp. 36-37.

23. Ibid.

24. Lummus is a 75-year-old U.S. engineering company based in Bloomfield, N.J., an affiliate of Combustion Engineering. Vepica is a 10-year-old Venezuelan engineering company based in Caracas.

25. *El Nacional,* March 19, 1979 p. A-4.

26. *The Daily Journal,* March 23, 1979, p. 5.

27. *Resumen* 341, April 1981.

28. *El Nacional,* March 15, 1980, p. D-8.

29. *Autentico* 90, March 5, 1979.

30. *El Nacional* March 3, 1979, p. A-4.

31. Anibal Martínez is a Venezuelan geologist, graduated from the Universidad Central de Venezuela, with a master's degree from Stanford. He worked for Creole for several years, resigned and joined the technical staff of OPEC in Vienna. Back in Venezuela, he has remained outside the oil industry as an independent critic. He is the author of several books on Venezuelan oil, including the valuable *Chronology of Venezuelan Oil* (London: Allen and Unwin, 1969).

32. *El Nacional,* March 21, 1979, p. D-5.

33. Dr. H. Peñaloza became a member of PDVSA's Board in 1979.

34. *Ultimas Noticias,* April 20, 1981, p. 10.

35. This would take place in September, 1982.

36. G. Coronel, in *Resumen* 430, January 1982, p. 10.

11 The Seeds of Failure, 1981

During 1979 and 1980 the performance of the oil industry was reasonably good. Income was up, production normal, projects in progress. Less favorable, however, was the industry's mood. Technical problems were mounting, the scarcity of experienced human resources was being felt acutely, operating costs were increasing, and, above all, outside pressures were being felt at all levels of the organization.

The president of Petróleos de Venezuela, who had for several years been the most powerful man in the oil hierarchy, was rapidly losing that leadership to Oil Minister Calderón Berti. As a result, Petróleos de Venezuela had started to look still more like a government agency and less like an autonomous state corporation.

This shift in power had been due to several reasons. Since August 1979 the board of PDVSA included members who had close personal ties with the minister and consistently tended to follow his lead. There was little doubt that the composition of the new board had weakened the position of PDVSA's president, General Alfonzo Ravard. Second, a group of top executives of the operating companies resented the fact that General Alfonzo Ravard was still the president, since that seemed to block their own advancement. Although they were not hostile, they no longer actively supported him. At the same time, Calderón Berti, contrary to the self-effacing previous minister, Valentin Hernandez, was determined to exercise his power and to increase the control of the ministry over the industry. He encouraged the exercise of authority by the ministry staff. Moreover Calderón Berti's personality helped him to become the perceived leader. He was an extrovert, liked to travel extensively to the operational areas, and loved to talk, not only to the top managers but also to the younger staff. In contrast, Rafaél Alfonzo Ravard was shy and aloof. He rarely traveled to the operational districts and felt uneasy surrounded by operators.

Some segments of the organization of operating companies such as the production divisions of Meneven in eastern and western Venezuela had for several years been under the supervision of engineers and managers who either had worked for the ministry or had developed close personal friendships with top ministry officials. This often led to direct contacts between Meneven and the ministry for the making of technical or managerial decisions over which Meneven should have consulted with PDVSA. Such an unusual liaison tended to undermine the power of PDVSA vis-a-vis the ministry.

Calderón Berti was enjoying some of his best moments in the international oil scene as well as president of OPEC. This, combined with the colorless nature of the rest of President Herrera's cabinet, had given him considerable political prestige. In January 1981 Mexico and Venezuela had joined in a regional aid program through which each country would contribute to the supply of crude oil and products to central American countries. This agreement was very favorable to Venezuela, and Calderón Berti deserved much credit for the way negotiations were carried out. President Herrera's total trust in his minister helped Calderón Berti to consolidate his influence on the industry. Moreover, sectors of the Caracas press started to eulogize him. Regular newspaper columns such as Asdrubal Zurita's "Tómbola" in El Mundo, Lossada Rondon's "Miraflores ä la Vista" in El Nacional, and E. Pérez Mirabal's column in 2001 systematically praised the young minister. Some of the praise was justified; some perhaps exaggerated.

Charges of Waste Are Made against the Oil Industry

The growing breach between the oil industry and the political and governmental sectors was dramatically illustrated in April 1981 when the president of Acción Democrática, the highly respected political leader Gonzalo Barrios, gave an interview to several Venezuelan newspapers in which he claimed that there was an unfair distribution of income in Venezuela and cited, as an example, "the fabulous salaries earned by the top executives of the state oil companies."[1] Barrios claimed to be under the impression that within the petroleum sector expenditures were "extravagant" (dispendiosos) and in style "closer to what prevailed in the lives of capitalistic tycoons." He went on to say that the National Congress should investigate these remuneration levels, since the oil-industry executives were state employees. Barrios added that he did not believe that political maneuvering or corruption existed in the administration of the petroleum industry and emphasized that he was only asking for information.

Although the opinions advanced by Barrios had the moderation which had become his trademark, they had a significant impact on both the oil industry and the political sector. It was the first time that one of the staunchest defenders of the autonomy of the oil industry had taken to the press to make a point of criticism and to ask for information that he could have asked for in private. More than what he said, it was the vehicle he utilized to say it that caused the reaction.

The chief whip of COPEI's congressional group, Oswaldo Alvarez Paz, a member of the Caldera group, replied immediately that "The oil industry has sufficient controls which guarantee its efficiency."[2] He admitted, however, that any abnormality in connection with expenditures should be promptly investigated. He warned: "We should take good care of that industry for the sake of present and future generations. I believe it is in good hands, but it is always useful to be alert." When asked if, in his opinion, Congress should increase control over the industry, Alvarez Paz replied: "I do not think so. Congress, through the committees, has closely followed the evolution of the oil industry and the management of the oil industry has always cooperated fully with Congress."[3]

Not every political leader agreed with Alvarez Paz. The president of the comptrolling committee of the Congress, Gustavo Mirabal Bustillos, said that Congress had been "waiting for 6 months to receive a list of the key managers of the oil industry, with their salaries" and that PDVSA had never answered.[4]

An unidentified executive from PDVSA, quoted in *El Universal,* made some critical comments abut Barrios's public outburst. He said, "Petróleos de Venezuela can answer in private to the executive committee of Acción Democrática . . . about the cost profiles of administering the oil industry, including the salaries of the high-level executives . . . (but) the oil industry does not deserve this type of treatment since there are many organisms of the public sector (which remain unmolested) which could properly be defined as wasteful."[5] With this statement, the battle lines between politicians and technocrats seemed fully drawn once more.

The opposition parties sided with Barrios. In *El Nacional* four leaders of the opposition expressed their views.[6] Julio Fuentes Serrano, president of the energy and mines committee of the lower Chamber of Congress and representative of URD said that "it was possible that many employees were being overpaid" and added that an investigation would soon start. Characteristically, however, he added that the investigation would only start "after the Easter holidays." Germán Lairet, from MAS said that "in effect, the salaries of the oil industry managers are too high and some projects, such as the Orinoco heavy-oil development, could be too costly to the country. We support an inquiry which would extend into the administration of the industry and the orientation of the existing petroleum policy." Siuberto Martínez, a MEP spokesman, denounced "a tendency in Petróleos de Venezuela and affiliates to avoid the control of parliament (which) should not be tolerated." José Vicente Rangel said that the oil-industry administration should be investigated since he had the definite impression that it was "costing the country too much." The leaders of the opposition parties were not only in favor of an inquiry but had already formed some strong opinions as to what the inquiry would find.

The vice-president of PDVSA, Julio César Arreaza, acting-president while General Alfonzo Ravard was absent, met with Dr. Barrios and provided him with information about oil-industry salaries. Arreaza told *El Mundo* that "only 6.9 percent of total expenditures of PDVSA goes to pay salaries for all staff" and added that PDVSA had given Barrios all the main documents relating to the plans and programs of the industry.[7]

An editorial of *El Diario de Caracas* advanced the following warning:

> Petróleos de Venezuela has been, up to now, one of the last apolitical strongholds. . . . Petróleos de Venezuela does not work for COPEI nor for AD, the right or the left. . . . PDVSA manages a national, nonrenewable resource. . . . In a corporation such as PDSVA there are bound to be short-comings and wrong decisions are probably made every day, but these are shortcomings which will exist in any corporation in the world. . . . When Barrios complains about high salaries and improper advisory fees, he does a disservice to a group which up to now has managed, with efficiency and dedication, the Venezuelan petroleum industry.[8]

The Venezuelan Congress decided to hold an inquiry in order to verify the Barrios's accusations. Octavio Lepage, president of the energy and mines committee of the Senate and prominent member of Acción Democrática, commented, "Dr. Barrios is usually very careful with words and if he said what he said, it must be because he has solid information that proves waste in the oil industry."[9]

Gonzalo Barrios came back on record to say that he was not satisfied with the explanations given to him by Arreaza.[10] He claimed that his interest was not so much in the salaries of the executives, but in the great projects of the industry. He added that in his opinion "it was not useful or convenient to disclose to the general public the list of salaries of high oil-industry executives." This was a confusing turn of events. Barrios now seemed to be sorry to have started what was shaping up as a major fracas. It was too late, however.

At this point a Caracas newspaper carried a piece of news to the effect that "already before the accusations of Dr. Barrios, the Minister of Energy and Mines, in a meeting with PDVSA, had warned them about the need to avoid temptations and excessive expenditures."[11]

José A. Ciliberto, the secretary of Acción Democrática's executive committee, continued the attack:

> We have information about the wasteful manner in which the oil industry is being managed. . . . We are dealing here with the squandering of many million bolívares per year. . . . We are guided only by the honest pre-occupation that the main source of income of Venezuela remains forever free from inefficiency and waste.[12]

The congressional inquiries had not even started and condemnatory judgment was already being passed. The impact of this attack on the top management of the oil industry was highly demoralizing and prompted me as executive vice-president of Menevent to comment:

> When the claim is made that the salaries of the oil-industry managers are too high, without knowledge of how these salaries are calculated, the country is not being helped. It will always be possible to find personnel who can be paid less, but perhaps these are not the personnel we need. The public administration is full of poorly paid employees, but this does not mean that we should imitate that example.
>
> . . . We refuse to heed the warnings of the petty politicians but invite the dialogue or the controversy with the better members of the political world . . . We believe in our mission. We are well paid, and pilfering is not our business. We are never absent from work, as is often the case with our critics. [13]

These opinions did not help to make the political sector feel any better.

The Congressional Inquiry

When the congressional inquiry finally took place, the 5-hour meeting dissolved into a surprisingly superficial exchange, which moved Jaime Lusinchi, secretary-general of Acción Democrática, to comment, "This inquiry has turned into a rhetorical contest sacrificing the original intention to force the minister to answer specific questions." [14] One senator appeared quite satisfied with the results of the inquiry, however. Senator Valmore Acevedo, of COPEI, said that there seemed to be no basic differences between government and the opposition, since Minister Calderón Berti seemed to be even stronger than Barrios in his plea for austerity. [15]

During the inquiry Minister Calderón Berti generally made a very strong defense of the industry and put most of its harsh critics on the defensive. He also said that "both in the meetings of PDVSA and in a private meeting held in Cabimas with the top management of the industry (he read the minutes of the Cabimas meeting), I recommended that top management adopt a careful and austere behavior, both in their activities and in social life. . . . I also advised them not to be arrogant." [16] This statement was deeply resented by the oil-industry managers for two main reasons. First, Calderón Berti implied that the charges of waste and extravagent expenditure leveled at the industry by the political sectors were true, and that he himself had become the conscience of the oil-industry executives regarding these practices. And second, he violated a gentleman's agreement made at the Cabimas meeting, on the basis of which it had been decided that none of the discussions held in this private meeting would be made public.

To appreciate the resentment of the oil-industry managers, one has to understand that the oil-industry executives *did* travel very frequently, always first class, often using corporate planes for those travels. They usually went to the best hotels and probably did not pay strict attention to travel expenses. Although this was true, it could also be argued that men who made stressful decisions involving millions of dollars had earned a high level of comfort. This was a strongly subjective and emotional issue. In the desire for comfort in travel, Calderón Berti was not excluded. To the contrary, he held a permanent vacation house in the beautiful oil camp of Puerto La Cruz, fully furnished and maintained by Meneven, one of the operating companies. Whenever he traveled to Puerto La Cruz, he would ask Meneven to have the refrigerator stocked with supplies and would order the company yacht retained for his family use. Calderón Berti traveled in oil-company aircraft until 1981, when he ordered the industry to buy a corporate jet, a Falcon-50, for his use as minister. He was the first Venezuelan minister of Energy and Mines who requested (and obtained) that the government pay the rent for his house.

The violation of the confidential minutes of the Cabimas meeting was of even greater significance. Most of the presidents and vice-presidents of the operating companies, as well as the board of Petróleos de Venezuela and some other high officers of PDVSA, had been present at that meeting. Calderón Berti had attended accompanied by his top advisors. Total confidentiality of the discussions held in this typical organizational development meeting had been agreed. When a few weeks later, the minister read minutes of that meeting not only in Congress but also during a television program, his credibility plummeted within the oil industry. His actions were seen by the oil technocrats as a political survival move.

Calderón Berti seemed to be moving away from the oil-industry camp and into the political camp. Gonzalo Barrios confirmed this when he said:

> Minister Calderón Berti himself seems to share my impressions, since in a certain private meeting held in Cabimas last February, he asked PDVSA to exercise moderation in their expenditures. . . . I did not know about the minister's statements until last night, but, according to Senator Acevedo Amaya, my observations were extremely mild as compared to the observations made by the minister.[17]

Calderón Berti's behavior during these events demonstrated that, if hard-pressed, he would not hesitate abandoning the side of oil-industry management. It was at this point that many people in the industry lost confidence in him.

The inquiry, which extended for several days, did not throw any light on the merits of Barrios's claims. Nonetheless it gave many opposition leaders the chance to voice their concern about certain areas of activity of the industry which, they felt, were not doing particularly well. The main areas of concern were alleged secrecy on the part of the industry, the Orinoco heavy-oil projects, and the increasing operating costs. The industry's "secrecy" was widely criticized by members of Congress. They felt they had a right to know all details about the industry, from the salaries of industry managers to the fine print of oil-sale contracts.

Regarding the Orinoco heavy-oil projects, worries centered on costs and on which foreign engineering company would get the contracts. To some congressmen like P. Márquez, "Bechtel, Lummus and Fluor (were) the new owners of our industry after Exxon and Shell departed."[18] In concern over the increasing operating costs in general, congressmen were joined by the oil-industry managers, who saw production costs increase every year without being able to do much about it. Inflation, rising labor costs due to government actions, and the obligation of the industry to buy local goods and equipment whenever possible, even at higher prices, had combined to increase operating costs from Bs 4.2 billion in 1976 to Bs 11.1 billion in 1980. P. Reimpell, director of finance in PDVSA, explained that this increase was due to inflation, salary raises ordered by the government, labor costs derived from the new labor-collective contract, and the increased activity of the industry in exploration, production, and all other aspects of industry operations.[19]

General Alfonzo Ravard made a detailed presentation in which he told congressmen that the industry could not work well without the support of the National Executive and the political parties and that this support would be best expressed by letting the industry operate without restrictions or excessive control. He added that PDVSA did not have secrets, but many of their activities had to be of a confidential nature because the oil industry was highly competitive.[20]

The oil industry came out of this inquiry largely untouched. *Zeta*, a Caracas weekly, made an analysis of the results of the congressional hearings and concluded:

The accusations about excessive expenditures (in the nationalized oil industry) do not seem well founded. Even if it were true that the oil-industry professional staff received the highest salaries in the country, this would be a fair remuneration to their competence. . . . The public opinion seems to feel that oilmen are right in desiring to keep their industry free from political interference, but they cannot ask that it be kept from policy-making. . . . The oil industry is still more efficient than any other state enterprise and, probably, than any private enterprise. There appears to be a natural resentment in the oil-industry managers when they are accused of being wasteful, while other state-owned companies lose enormous amounts of money and when criticism comes from a Congress which is characterized by laziness and inefficiency.[21]

Why don't congressmen investigate the state-owned enterprises which suffer such enormous losses? asked *Zeta*. Not a bad question, but one which remained essentially unanswered.

The Events of August 1981 and the New Board of PDVSA

The approach of August 1981 meant that a new PDVSA board had to be named. Rumors, pressures, lobbying, and internal strife were more evident than ever before.

August 26, three days before the date of the new board of PDVSA was to be named, the shareholders met and made other important decisions. Minister Calderón Berti called a press conference, after the meeting, to announce that the shareholders "had decided to move the main offices of Meneven away from Caracas to Barcelona–Puerto La Cruz and of Pequiven from Caracas to Maracaibo." This decision had a dramatic impact on many top oil-industry managers and caused an almost open rift between operating-company management and the minister. It needs to be analyzed in detail.

How the Decision Was Made

Usually administrative decisions of this magnitude are the object of considerable analysis within an organization, not only concerning their political and organizational impact, but also regarding their logistics and economics. Moving the headquarters of a large corporation from one city to another in the United States is almost always a complex decision, as the moving of several large oil-company headquarters into Houston several years ago illustrated. In Venezuela such a decision was momentous, since Caracas is not only the largest city in Venezuela but the site of government and the place where all important decision-making takes place. To be away from Caracas means, unfortunately, to be away from the mainstream of events. No large Venezuelan corporation could survive for long without a strong presence in Caracas. For these reasons it should have been imperative to conduct a careful analysis of the pros and cons of moving Meneven and Pequiven away from Caracas.

This analysis was never made—not by the ministry, where the decision originated, and not by PDVSA, Meneven, or Pequiven. Furthermore the operating companies Meneven and Pequiven were never asked to produce such an analysis, nor were they formally asked for their views on the moves.

The only documented official action in this connection took place in January 1981, at the PDVSA shareholders' meeting, when Minister Calderón Berti asked PDVSA to conduct a feasibility study of the possibility of moving Meneven to Puerto La Cruz–Barcelona. This study was assigned to three directors of PDVSA but 5 months later, as was often the case in PDVSA, the study had not even been started. No doubt angry and frustrated, Calderón Berti started talking to the press about the moves. He said in June, "The government is studying the possibility of moving away from Caracas the headquarters of the operating companies of the oil industry."[22] It is significant that Calderón Berti said "the government" and not PDVSA. This was an indication that the decision being contemplated was essentially political.

Traveling to Barcelona, Calderón Berti also spoke to the regional press about the moves. Reported *El Nacional:*

> The minister said it was probable that Meneven will be moved to eastern Venezuela. . . . such a move would be convenient to the interests of this area. It would stimulate economic development and it would contribute to the progress of this area. . . . it would be very important to consolidate the activities of Meneven specifically in this area.[23]

Calderón Berti stated in Puerto La Cruz: "It is almost inevitable the move of operating oil companies (away from Caracas)." He added that he "definitely favored the move of Meneven to Puerto La Cruz–Barcelona."[24] All of these press statements were made long before a decision had been announced and before PDVSA or the companies potentially affected by the moves knew anything official. A senior oil-industry executive would comment: "It got to the point that we began to feel that decision-making in the oil industry was being done by the minister and informed to us through the press. This was indeed a new experience."[25]

That the decision to move Meneven's headquarters to the interior of Venezuela was essentially a political one seemed to be confirmed by the governor of the State of Anzoategui (where Puerto La Cruz and Barcelona are located). Governor Guillermo Alvarez Bajares, stated,

> This decision of the Venezuelan government (to move Meneven to Puerto La Cruz–Barcelona) announced by Minister Calderón Berti and PDVSA is the fulfillment of a promise made to us by President Herrera Campíns. . . . Minister Calderón Berti, the spokesman for the government decision, deserves much credit for the personal interest he took in the decision which will lead to the move.[26]

Alvarez Bajares thus clearly admitted that this decision had been taken by the government at the highest political level, much before the oil industry had been involved, and that Minister Calderón Berti was deeply involved.

The frustration that Minister Calderón Berti might have rightly felt at PDVSA's rather negligent handling of the request he had made for a study in January 1981 did not seem enough to justify his unilateral exercise of authority. In fact it now appears probable that PDVSA's negligence might have served to lend unexpected support to Calderón Berti's essentially political decision. What seemed to be inconsistent on the minister's part was to ask for a feasibility study and then to make the decision without having it.

The Meneven and Pequiven boards did not find out about the decisions until several hours after they had been made—in fact after they had been announced to the press by Minister Calderón Berti.[27]

The way these decisions had been made represented a major departure from traditional decision-making processes in the Venezuelan oil industry. In 1976 it would have been unthinkable that such a thing would happen. In 1979 it would have been improbable and hard to accept. In 1981 it had happened.

The only reaction of significance came from me, as vice-president of Meneven. I sent Alfonzo Ravard the following letter:

August 1981

General R. Alfonzo Ravard
President
Petróleos de Venezuela
Caracas

Dear General:

As you will understand, the recent decisions about Meneven and the way they were taken, make it very difficult, practically impossible, for me to remain on the Board of this company. However, after 26 continuous years of service to the industry, I would wish to remain on it. Therefore, I am formally asking to be transferred to another operating company as soon as possible. Maraven preferably, or Lagoven, no matter the place or the level of the assignment. I am willing to accept any reasonable managerial or technical job where I could be of use.

I will be back in Venezuela September 10. Good luck to you in these coming two years.

Greetings

G. Coronel[28]

This letter was taken by PDVSA as a letter of resignation from the industry. In October 1981 I was severed from my job without even being able to meet with PDVSA officials or with the minister. This style of dismissal of top oil executives was new to the industry, where decisions of this nature had never been made without a personal interview with the executives involved.

The New Board of PDVSA

President Luis Herrera Campíns finally announced the names of the members of the new board in the first week of September 1981. This announcement came late, one week after it was due. This seemed to indicate that the decisions had not been easy, certainly that no systematic, meritocratic procedures had been applied.

Hugo Finol, perhaps the most knowledgeable of the members of the board in the area of production, was removed without any explanation, public or private. Again, in this case, no interview with the ousted executive had taken place. Finol's fault had been, according to the oil industry version, to disagree with the views of the minister in production matters. Sources close to the minister claimed that Finol had been held responsible for not completing the feasibility study of the Meneven and Pequiven's moves. The dismissal of Finol was publicly protested by AGROPET in a press communique, which said, among other things:

> We feel that we have earned the right to ask that the selection of the top executives of the industry should be made on the basis of meritocracy and the quality of the career of the candidates. . . .

> What AGROPET considers inconvenient is that . . . the press should become the means to tell a top executive that he is being removed as was the case (with Finol).

> The fact that an executive such as Dr. Finol, with an impeccable career of more than 20 years . . . reads about his dismissal from the board of PDVSA in the press makes us doubt about our own stability (as oil-industry employees).[29]

Because of the pressure exerted on the minister by top oil-industry executives, notably those at Maraven, Finol finally found a place at the board of Lagoven. Otherwise his career would have ended abruptly, in open violation of all the rules and norms of personnel administration existing in the oil industry.

General Rafaél Alfonzo Ravard was, once more, retained as president, Julio César Arreaza as vice-president, and Victor Petzall was promoted to second vice-president. The permanence of General Alfonzo Ravard for the new period 1981–1983 had still deeper significance than his having been retained for the period 1979–1981. First, it seemed to destroy once and for all the expectations of the top oil executives who had been in line for that job and consequently changed permanently one of the original rules of the game. Rodriguez Eraso now seemed to be out of the race since he would reach retirement age sometime in 1983. Quirós was being forced to wait 2 more years, and his patience logically was wearing thin. The worst aspect of the situation seemed to be that there was no longer any solid basis to expect that in 1983 a promotion from within the industry would take place. The line seemed to have been ruptured.

The perception grew that General Alfonzo Ravard was being used as a pawn in the power play being enacted by Minister Calderón Berti. In December 1983 there would be national elections for president. The probabilities were high that Acción Democrática would win those elections, reinforcing a trend already established in the last 15 years. Whatever the electoral outcome, Minister Calderón Berti seemed to be trying to assure himself of a prominent position in Venezuelan political life. He apparently had chosen the presidency of Petróleos de Venezuela as his next niche. In August 1983 President Herrera and his oil minister (Calderón Berti) had to agree on a new board for Petróleos de Venezuela, and this board would remain in control until August 1985. Such a period as president of Petróleos de Venezuela could considerably strengthen Calderón Berti's image and reputation and prepare him for his next possible step, somewhere in 1988, or even, in 1993: the presidency of Venezuela. Minister Calderón Berti was young and he could wait. Venezuelan political rules, in fact, dictated that he should wait. Being at the helm of the most powerful economic organism of the country did not seem too bad a place to wait. Venezuela was an oil country. A man still in his prime, who had already been the political, managerial, and economic leader of the oil industry would no doubt seem to be a natural choice for even higher positions.

The permanence of General Alfonzo Ravard at the top job of PDVSA for the period 1981–1983 was almost mandatory under this scenario. He was already past his retirement age. He posed no long-term threat to Calderón Berti and would be more than glad to be able to say that he had been chosen, over and over again, for this prestigious job until the time came for him to decide that he did not want to continue. Naming an oil-industry executive as president of PDVSA would have made the success of Calderón Berti's gambit much less plausible, almost impossible, as traditional industry values would have been so strongly reinforced that another future disruption would have been extremely unlikely. Furthermore the permanence of Alfonzo Ravard in the job assured Calderón Berti that he would have no real opposition as the top industrial leader, since Alfonzo Ravard clearly understood that the price he had to pay for survival was the almost total surrender of power. The naming of Rafaél Alfonzo Ravard as president of Petróleos de Venezuela for the period 1981–1983 was therefore a political act. Alfonzo Ravard obtained what he wanted; Calderón Berti obtained what he wanted; the top oil-industry hierarchy decided to wait. For them 2 years was a period of time in which many things could happen and in which the old story of the prisoner who gets a commuted sentence could come true.[30]

In doing so, however, these main actors were tacitly accepting the new rules of the game: political skill had now become the passport to power in the Venezuelan oil industry, something not true before. The argument that a minister should not become president of PDVSA because this would replace normal career progression by political shortcut apparently was no longer valid.

The promotion of W. Petzall to the second vice-presidency clearly gave Calderón Berti a better grip on the board. This time it had apparently proved to be indispensable to retain Arreaza as first vice-president, in order to keep Acción Democrática resonably satisfied. Petzall, however, would become Calderón Berti's key man on the board.

Some of the new members of the board did not seem to have a solid claim to the job when compared with many other oil-industry executives. In particular the naming of E. Daboín and F. Guédez was surprising. E. Daboín, chosen a principal director, was a petroleum engineer with a working experience of about 21 years. Most of his experience, however, had been as technical inspector at the Ministry of Energy and Mines, mostly concerned with the auditing of operations in the oil fields of eastern Venezuela. He had little or no managerial experience and no exposure to project management, to planning or to the tasks of manpower development and organizational analysis which were the areas in which he would have to be involved at PDVSA. Daboín had always been a middle-level supervisor with a limited grasp of the more complex corporate environment. His selection as member of PDVSA's board was indubitably prompted by his very close personal ties to minister Calderón Berti, who was his boss at the ministry and who would still be his boss at PDVSA. In selecting Daboín, the government seemed to break away from the concept of meritocracy.

F. Guédez, chosen as alternte director, was also a petroleum engineer, with more than 20 years' experience. He had a much stronger managerial background than Daboín, since having been general manager of the eastern Venezuela production district of Meneven, and later, a member of Meneven's board. He was ambitious, a hard worker, and had leadership qualities which, if properly utilized, could make him become a good executive. However, he did not seem to be ready for the job at PDVSA. Some months before, he had been turned down, both by Lagoven and Maraven, as a candidate to sit on the board of those companies, on the grounds that there were much stronger candidates within the industry. It was illogical, therefore, that he should be now considered ready for a seat at the board of the holding company. Here again personal connections prevailed. Guédez had been, from his position as Meneven's production director, one of the staunchest allies of the Ministry of Energy and Mines in the fight for power between the ministry and PDVSA. Guédez had done all he could to reinforce the authority of the ministry, sometimes at the expense of the authority of PDVSA. As a result the quality of the relationship between Meneven and PDVSA had deeply deteriorated. Once at the board of PDVSA, therefore, there was little doubt that Guédez would strongly reinforce the minister's camp.

The question which inevitably arises is why PDVSA accepted these two candidates as new board members. The answer seems to be because they had no choice. General Alfonzo was again fighting for survival, and he simply was not strong enough to block these nominations. He had already vetoed the nomination of Arévalo Reyes to replace Arreaza as vice-president. Although Reyes was also an Acción Democrática man, he was at such odds with the industry and with Alfonzo Ravard that his nomination was blocked by Alfonzo as the one candidate he would not accept. In taking this stand, however, Alfonzo Ravard had to take Daboín and Guédez.

The only other new member of the board was Nelson Vazquez, an engineer of considerable experience in the industry, who had been up to that moment president of INTEVEP, the research and development company of PDVSA. Vázquez's choice was not objectionable since he was a solid manager with sufficient exposure at the highest corporate levels. He was also a follower of Calderón Berti but apparently not unconditionally so.

Minister Calderón Berti said, in reference to this new board:

> It has been formed as the result of an analysis made by President Herrera Campíns and the (cabinet). The team will blend the experience of the old members with the contributions of the new members. . . . No one of the new members has less than 22 years' experience in the petroleum sector, which invalidates any idea that we are improvising in the selection of the top personnel of the industry.[31]

Obviously the minister knew that there would be comments about the choices made and wanted to take the initiative in disqualifying such comments before they were made.

A few weeks after all these unusual events had taken place, General Rafaél Alfonzo Ravard gave an extensive interview to a Caracas newspaper in which he said: "Justice and honesty are always present in our dealings with human resources within the oil industry." General Alfonzo also said, in reference to his reelection for 2 more years:

> I have understood my commitment in a wider sense than (simply) with a company. . . . I have said many times that it is fundamental for the success of the managing of the oil industry to remain loyal to the technical and professional administrative mechanisms of managers. . . .

> During all these years we have maintained the objective of keeping the oil industry from being politicized. The careers of the men who make up the industry do not depend on the changing circumstances of partisan politics. Their recruiting and promotions depend on the systems of personnel administration (which are, in turn) based on individual merit as determined

by the supervisors. . . . This scheme allows for the promotion to the highest
levels when circumstances and personal merit justify it.

When asked by the journalist, "What do you think of the comments which
suggest that Minister Calderón Berti influenced directly the selection of the
board of PDVSA and that, now, he has 'his' people in this board?", General Alfonzo answered, very noncommitally: "The by-laws (of Petróleos de
Venezuela) give the president of the republic the right to name the board of
Petróleos de Venezuela. This right allows him to establish the mechanisms
of advice that he desires." According to this interview General Alfonzo felt
that all that had taken place at PDVSA during the last 2 months, including
the dismissal of H. Finol, the decisions to move the headquarters of Meneven and Pequiven away from Caracas, and the naming of Daboín and
Guédez to the new board of PDVSA, had occurred strictly within the existing norms and procedures of the industry and in accordance with "justice
and honesty." As the reelected president of PDVSA, General Alfonzo certainly could not say anything else. Perhaps he could have chosen to remain
silent instead, since this time there was a strong perception in many quarters
that he had paid a high price for survival. After all, one of his favorite sayings had always been: "We don't always have to say all of the truth, but all
that we choose to say must be true."

The Reactions Against Politicization

Protests against these events came from several quarters, although there
was no open conflict. Journalists Kim Fuad and Gerardo Inchausti, writing
in a Caracas weekly said,

> There is an increasing perception among veteran oil-industry personnel that
> the "hands off" policy which characterized the first 5 years of the (operations of the) state-owned oil enterprise is no longer being applied. The last
> changes at board level in PDVSA and the operating companies have
> prompted a reaction of displeasure since political considerations have been
> introduced in decisions that traditionally have been governed by strict
> managerial considerations.
>
> The (oil) industry feels that its capacity to respond is being weakened by
> pressures from government and the political parties which, in turn, tends to
> reduce its capacity for professional and responsible decision-making. The
> industry, whether it likes it or not, is becoming subordinate to the political
> world. . . . An example of the gap between the two worlds came in September when CORDIPLAN (the planning organism for the state) Minister
> Ricardo Martínez suggested that PDVSA's investment fund—some Bs 35
> billion—could be utilized in expenditures not related to the petroleum sector. . . .

The demands being made on the industry are not limited to the use of local goods and services but also include the support to inefficient state-owned enterprises and the direct financing of nonpetroleum projects and activities. This is in conflict with the original objectives of the nationalized (oil) industry.

Fuad and Inchausti were not alone in their worries. The *Journal of Commerce* of January 26, 1982 ran an article in its front page signed by John Sweeney, in which the author claimed that "the Venezuelan oil industry is being forced to subordinate its objectives to the political requirements of the government of Luis Herrera Campíns." He added that "a majority of the decisions in the oil sector are being taken by the government of Luis Herrera Campíns without consulting with PDVSA or its affiliates."[34] The article also claimed that the Venezuelan petroleum industry was being forced to finance roads, hospitals, and schools.

The *Wall Street Journal* reported:

Petróleos de Venezuela's achievements in the country's highly charged atmosphere have apparently only increased its attractiveness as a political plum to be plucked. So, forgetting the original vow to keep hands off, President Luis Herrera's government has launched a full-scale intervention in the company's affairs. Oil Minister Humberto Calderón Berti calls it "the deepening of nationalization." To Alberto Quirós, who heads one of Petróleos de Venezuela's four operating subsidiaries, it means that "the honeymoon is over."

Mr. Calderón Berti further jolted Petróleos de Venezuela management by naming members of the firm's board of directors who were personally loyal to him. Company insiders say the new members couldn't have met the usual criteria of merit and experience that usually govern who goes on the board. They express fears that Mr. Calderón Berti's action threatens the company's carefully nurtured system of promoting on merit rather than on political credentials.

The Company is also under pressure to buy Venezuelan-made oil-field supplies rather than seeking equipment on the basis of quality and price . . . The beleaguered oil company doesn't even find defenders among the government's political opponents. Gonzalo Barrios, president of the rival Democratic Action party, has charged that Petróleos de Venezuela "is engaged in extravagant expenditures and its executives act more like barons of a multinational company than executives of a state company."

The report ended:

Company executives privately wonder how long standards will withstand the attacks of the politicians. The ousted Mr. Coronel . . . compares Petróleos de Venezuela to the New York Yankees' baseball team and Oil Minister Calderón Berti to Yankee principal owner, George Steinbrenner. "The Yankees are the richest team with the best players," he says, but they "have the greatest organizational problems because the owner wants to run the team from the stands."[35]

Abelardo Raidi, a very well-known and well-read Venezuelan journalist, known to be a COPEI sympathizer, had this to say:

> Rumors are on the increase concerning the displeasure of top oil-industry management. Apparently the natural leaders of industry accepted, at the moment of nationalization, to stay at their jobs in the operating companies for a transition period (the first 4 years) in order to ensure an orderly change. . . . However, the second board of PDVSA already contained a political ingredient. The third board reinforced this tendency since, by the first time in industry, there were appointments based on friendship and personal loyalties and not on the meritocratic planning system which had characterized (promotions in) the oil industry. . . . The problem is that, if this trend continues, there is a real risk that oil employees start to play politics in order to be promoted, forgetting the rules of the game which have prevailed for so many years.[36]

Business consultant Andrés de Chene said, in his Management Letter of February 1982: "The criticism against Calderón Berti is that he has initiated the politicization of Petróleos de Venezuela by making personal appointments who are not qualified. . . . They require external help to remain in those jobs. Has anyone ever heard that (competent) people need a godfather to be promoted?" Perhaps the most significant reaction came from Alberto Quirós, president of Maraven. In a very important series of articles written for the Caracas daily, *El Nacional,* from September 1981 onward, Quirós strongly criticized the politicization of state-owned industries and generally described the differences between the political and the technocrat worlds. The fact that he never mentioned the oil industry or any person in particular was no obstacle for his readers to understand clearly what he meant to say. Quirós acted very courageously, as he had much to lose. In the absence of open support from his more cautious colleagues at the highest management levels, he became a somewhat lonely but formidable figure of dissent. He said,

> We might think we have been able to persuade when we only have forced silence upon the adversary. We tend to see agreement where there is only absence of open dissent. . . .

> In an efficient organization . . . to have the last word requires that, before it is said, many other words should have been spoken. The last word is usually said by someone with considerable experience, by a carefully selected leader. Some organizations have developed a system in which the right to have the last word is obtained by decree and not through personal merit, where the position is more important than the person and . . . where dissent is taken as insubordination. Unfortunately, this is the case . . . of almost all of our state-owned enterprises.

He would add, "These organizations lack identity, lack objectives and are made up by professional survivors who adjust their thinking to the wishes of the boss." And: "Politicization consists in . . . trying to apply to technocracy the rules of politics . . . the essence of progress and democracy lies in the defense of our right to dissent . . . in the managing of conflict and not in its suppression."

Quirós also said,

> Politicians are elected. Technocrats are selected. . . . Personal loyalties can make the difference between success and failure in the political world. . . . Therefore, the politician chooses his personnel on the basis of loyalty. This is why, in the political world, there exist nepotism, cronyism, appointments by friendship. . . . Technocrats are different. Selection is made on the basis of ability, competence. It is almost impersonal. Technocrats do not have friends, brothers, or cronies.

In a special supplement to his managerial publication, "The Monthly Report," economic and political analyst R. Bottome stated:

> Venezuela's oil industry is fortunate to have many experienced, highly trained technicians and executives. However, some have already left while other are increasingly frustrated by the apparently irrational decisions forced on the industry by the state's bureaucrats. The gut issue is that PDVSA management is increasingly unable to act independently, and its managerial performance is being judged with criteria which have nothing to do with sound management priorities of efficiency, profitability, and long-term development goals.[37]

A short, harsh note in *Zeta* read:

> The inexpert and unconditional friends of Dr. Calderón Berti insist in helping him reach the presidency of PDVSA at the expense of the dismantling of our (oil) industry.
>
> Calderón Berti is being left alone in the ministry since the most qualified technicians do not wish to be coresponsible for the tragic mistakes that are being committed in the design and execution of our petroleum policy.[38]

And in a long interview given to Jorge Olavarría, editor of *Resumen,* I said, among other things:

> All previous Energy and Mines ministers were very dedicated to the job and completely disregarded any personal consideration or political ambition. This is not the case today. . . . The talent of the new minister . . . is not being specifically utilized in the improvement of the sector under his command, but is being utilized . . . to enhance his future political career and to create an enclave of personal power within the oil industry.[39]

The Venezuelan Oil Industry at the End of 1981

Some of the main economic indices of the petroleum industry during 1981 were very positive. Although production and exports went down 2.8 percent and 5 percent, respectively, these reductions were compensated by an increase of 13 percent in the average realization price of crudes and products, which reached $29.71 per barrel. National participation per barrel went up 9 percent to about $22.35. However, some other indices gave cause for worry.

> Local consumption of hydrocarbons grew by 7 percent. This was essentially due to the delays in the completion of electrical generation projects, the other main source of energy.

> In spite of the enormous effort in production drilling and repairs, the production potential did not increase significantly, staying at about 2.4 million barrels per day.

> The OPEC meeting, held in Geneva in October 1981, decided to unify the crude-oil prices of the number countries on the basis of $34 per barrel for the Arabian light marker of 34°AP and to establish price differentials on the basis of quality and geographical distribution. To adjust to this decision, the Venezuelan petroleum industry had to reduce the prices of its light crudes in about $2 per barrel as of November 1981.

> The market weakened noticeably in the second semester of 1981, and there were strong signs that this weakening would extend into 1982.

After the decision to reduce light crude-oil prices had been made to conform to OPEC's October resolutions, Minister Calderón Berti made a rather controversial decision. He ordered, as of November 1981, an average increase of about $0.60 per barrel in the export prices of the heavy crude oils and claimed that this increase would compensate for the loss of revenue derived from the reduction of the light crude-oil prices, giving the country an overall positive balance for 1982.

But reality was different, and the oil industry knew it. At the level of the operating companies, this increase was seen as a purely political maneuver designed to make the government bureaucracy look good after the October OPEC meeting. Oil experts knew that the market would not accept an increase and that such an increase would simply serve to weaken further the position of Venezuelan oil in the world markets. The measure was perceived as short-term oriented, ordered in the hope that some unforeseen event improved the market dramatically. This hope was not realized. In January 1982 the government had to order a reduction of $0.70 per barrel in the average price of heavy crudes.[40] After the unrealistic price increase of November, the government had to rectify.

A few days later Minister Calderón Berti still claimed, "In 1982, oil income will not decrease due to our policy of stable pricing."[41] By this time it seemed obvious that Minister Calderón Berti was particularly concerned with his long-term political future. After his return from the OPEC conference held in Bali, Indonesia, he called a press conference in which he said,

> It was a highly tense meeting. . . . I was always optimistic about the outcome, and our thesis . . . prevailed. An important factor (in the outcome) was the meeting I had with Professor Subroto (Indonesia) and René Ortiz (Ecuador) . . .

Talking about Iran, he continued:

> Hassan Sadat (of Iran) told me that the (American) hostages would be freed before year's end. We talked for some time in my quarters since he wanted to talk to me. . . . But I think it was in Kuwait and Paris where I could really persuade several of the members (to adopt my thesis)[42]

The Venezuelan newspapers described the Bali meeting as a triumph for Venezuela. It seemed to the local press that Minister Calderón Berti was always returning from these OPEC meetings victorious. When the October meeting forced a price reduction for light crude oils, he therefore felt compelled to raise heavy-oil prices so as to offset the perception that Venezuela had "lost." During the first 2 months of 1982, however, Venezuela would be forced to reduce oil prices three times.

By the end of February 1982, the country was already in the deepest economic crisis of the last 25 years. The oil market had suffered a collapse. Minister Calderón Berti nonetheless would still claim: "We are confident that by the second half of 1982, we can and we shall see an increase in the demand, with an immediate beneficial effect on prices."[43] The economic crisis which rocked the nation in 1982 was due to factors largely outside Venezuela's control. No oil-producing country except perhaps Saudi Arabia could influence decisively the price structure of oil in the world markets. Venezuela certainly could not. What is important to mention at this point is that this major crisis found an oil industry in which the seeds of failure had already been planted. Politicization, cronyism, personal ambition, and distortion of traditional values had somehow taken hold. As a result the capacity of the industry to respond rapidly and effectively to crisis had significantly decreased.

Notes

1. *El Carabobeño,* Valencia, April 8, 1981.

2. COPEI, in government, had also started to drift apart and by 1981, there were two very distinct COPEI groups, the "Calderistas," followers of Rafael Caldera, the founder of the party, and the "Herreristas," followers of President Herrera and the former secretary general of the party, Pedro Pablo Aguilar.

3. *El Universal,* April 9, 1981.

4. *El Universal,* April 9, 1981.

5. Ibid.

6. *El Nacional,* April 9, 1981.

7. *El Mundo,* April 10, 1981.

8. *El Diario de Caracas,* Editorial, April 10, 1981.

9. *2001,* April 10, 1981.

10. *El Nacional,* April 11, 1981.

11. *El Universal,* April 12, 1981, a column by Carlos Croes.

12. *El Nacional,* April 14, 1981.

13. *El Universal,* April 19, 1981, p. 1-12.

14. *El Universal,* April 23, 1981, p. 1-14.

15. *El Universal,* April 23, 1981, p. 1-14.

16. *El Nacional,* April 23, 1981, p. D-3.

17. *El Nacional,* April 24, 1981, p. D-2.

18. *El Diario de Caracas,* April 27, 1981, p. 9.

19. *El Universal,* April 28, 1981, p. 1-23.

20. Ibid.

21. *Zeta* 371, May 3, 1981.

22. *El Diario de Caracas,* June 25, 1981, p. 10.

23. *El Nacional,* June 6, 1981, p. D-4. News from Barcelona, reported by E. Marin and E. Lira.

24. *El Nacional,* August 24, 1981.

25. Personal interview with a member of the board of Meneven, May 1982.

26. *Ultimas Noticias,* September 1, 1981, p. 6.

27. As vice-president of Meneven and acting president at the time, I received a telephone call from a member of PDVSA's board at 7 P.M., 3 hours after the minister had given his press conference. Even this call was informal. The formal communication was conveyed to Meneven the morning after, at 10 A.M., by General Alfonzo Ravard.

28. In my files. This letter was copied to G. Rodriguez Eraso and A. Quirós and to J. Chacin, presidents of Lagoven, Maraven, and Meneven, companies mentioned in the letter.

29. *El Nacional,* September 7, 1981.

30. A prisoner sentenced to death accepted a 2-year postponement of his execution, offered to him by the king of the country on the condition that he teach the king's horse how to fly. His cellmate laughed at his decision, which he perceived as a cruel extension of his agony. The prisoner answered, "In two years many things can happen. The king might die and my sentence be suspended. I might die and the matter be resolved. The horse might die, in which case I could always claim that, had he lived, I would have been able to teach him how to fly. And of course, the possibility always exists that the horse might learn to fly."

31. *El Nacional,* September 4, 1981, p. D-8.

32. *2001,* September 14, 1981, p. 5.

33. *Número* 77, November 22, 1981, p. 15.

34. Quoted by *El Diario de Caracas,* January 27, 1982, p. 27.

35. The *Wall Street Journal,* February 16, 1982, p. 31. Reprinted by permission, © Dow Jones & Co, Inc. 1983.

36. *El Nacional,* December 24, 1981.

37. "The Monthly Report," special supplement, December 18, 1981.

38. *Zeta,* May 23, 1982.

39. *Resumen* 430, January 31, 1982, pp. 4–11.

40. *El Universal,* January 1, 1982, p. 2-2.

41. *Diario de Caracas,* January 7, 1982, p. 22.

42. *El Diario de Caracas,* December 21, 1980.

43. *El Universal,* February 26, 1982, p. 1-12.

12 Lean Years, 1982 and 1983

By using oil as an economic and political weapon, OPEC had become, during the 1970s, a formidable geopolitical power in an era in which the delicate military balance attained by the Communist and Western blocs already precluded the open use of force as a means of persuasion. But, silently and determinedly, oil consumers fought back. Combining decreased oil consumption and improved efficiency, the leading industrialized countries had managed to reduce their oil-import requirements by 1980. As compared to the 1960s, the United States managed a reduction of 7 percent in the use of energy per unit of gross national product, whereas in Japan this index dropped by 9 percent. For 1981 the overall oil consumption in the non-Communist world was down to 46.5 million barrels per day, about 5 million barrels per day lower than during the crisis of 1979 and 2.5 million barrels per day below the 1980 level. Projections of oil demand made in the United States suggested that by the year 2000 the country would be using 700,000 barrels per day less oil than in 1980.[1]

During the years of crisis of 1973 and 1974, OPEC had provided 60 percent of all world oil supplies. In 1981 this figure had been reduced to 40 percent as Alaska, the North Sea, and Mexico increased their share.[2] All in all, OPEC's production by 1981 was 27 percent lower than in the preceding year.

The year 1982 opened against this background of progressive softening of the world oil market. Oil-industry technocrats in Venezuela could see the crisis coming, but the government did not. A. Quirós, president of Maraven, stated that his company had detected well in advance what would happen and furthermore, that the crisis would be with the country for some time. He said,

> We noticed three things: first, that residuals (were) not our best product for future exports; second, that we had an abundance of heavy oils, and third, that these heavy oils should be dedicated to the making of special products such as lubricants or be subject to deep conversion in order to obtain marketable products. . . . Against this background we designed a marketing strategy emphasizing sales of heavy crude oils and the signing of long-term, stable contracts.[3]

219

As a result, Quirós added, Maraven's exports had not suffered as much in early 1982. Obviously, what Maraven had seen could have been seen by others as well. An important ingredient in the problems confronted by Venezuela during 1982 and 1983 was the failure of the government to read the situation correctly and to act in a decisive way.

By February 1982 the main export product of Venezuela, residual fuel oils, was not selling. The tanks contained about 25 million barrels of this product. Mexico had entered the U.S. residual market and was supplying around 200,000 barrels per day at prices significantly lower than Venezuela's.[4] For the third time in 2 months, Venezuela had to reduce residual fuel oil prices.[5] To many members of the Venezuelan political establishment the drop in oil consumption and the weakening of oil prices were simply "a maneuver of the industrialized countries to harm OPEC."[6]

A preliminary estimate of oil revenues for 1982 made by government analysts suggested that oil revenues would decrease by about $1.8 billion for the year, a drop of about 15 percent as compared to 1981, but oil-industry executives feared that the reduction would be significantly greater, perhaps as much as 35 percent. Petróleos de Venezuela decided to cut its operational costs by 10 percent, some $400 million, in an effort to minimize the overall deficit, but such a reduction carried the risk of limiting the production and the money-generating capacity of the industry.

By the end of February Minister Calderón Berti announced that the situation "would change favorably. . . . There will be an economic recovery in the second semester of 1982,"[7] but by this time few people in the country shared his view. Arturo Hernández Grisanti, Acción Democrática's oil spokesman, severely criticized the government for "irresponsible behavior and lack of planning" and predicted that the loss in oil income would be of $2.5 to $3 billion.[8] Political analyst L.E. Rey pointed out that, while the country was facing an imminent economic crisis, the government was still very intent on establishing a food-stamp program which would cost around $700 million per year and which would not significantly improve the living conditions of poor Venezuelans.[9] Oil expert and businessman Rafaél Tudela, a sympathizer of the party in government, felt compelled to warn that "The crisis is worse than we imagine it. Venezuela will have an income reduction of $4 to 4.5 billion."[10] An editorial, *El Diario de Caracas* cast doubt on the validity of the argument offered by President Herrera that the crisis could not have been predicted.[11] There was a clear divorce between the government's and public opinion concerning both the predictability and the extent of the economic crisis.

OPEC met in March 1982 to review the world market situation and concluded that oil production had to be curtailed in order to maintain prices. As a result a production ceiling of 17.5 million barrels per day was set; more important, each country was assigned a production quota. It was the first occasion that OPEC had behaved as a true cartel, and it came at a time in which severe measures of this type were needed in order to save the organization.

Saudi Arabia reduced production once more, by 500,000 barrels per day. Venezuela was given a quota of 1.5 million barrels per day. The market did not seem to react favorably to the move, however. Price remained weak and it looked as if the production reduction had not been sufficient.

The Government Begins to Eye PDVSA's Oil-Investment Fund

By the end of March members of the cabinet were beginning to consider the oil-industry investment fund as a possible source of financial reserves in case the situation worsened. In a private meeting with a group of Acción Democrática leaders, Minister Calderón Berti assured them that he would not allow this to happen. He described the crisis as "very temporary."[12] At this time, PDVSA made known its results for 1981, and its president R. Alfonzo Ravard claimed that "it would have been impossible (for anyone) to have been able to predict the change in the oil-market experienced in early 1982 on the basis of the 1981 market situation."[13] However, the same day Maraven's president, A. Quirós, in a talk to the Venezuelan-American chamber, said that the trend toward a softening of the demand and of prices could have been previously detected. Quirós claimed that the main problem facing the country was the extreme interdependence between Venezuelan public expenditure and oil income. What could and had to be controlled, he claimed, was public expenditure.[14]

Fedecámaras, the association of Venezuelan businessmen, warned against any government intentions of tapping the oil-investment fund for non-oil-related uses and stated that "anybody who decides to take (that) money away from PDVSA will have to bear a grave responsibility."[15]

In an effort to overcome the fiscal crisis and to instill confidence among the Venezuelan public, President Luis Herrera announced, April 9, a series of measures of a diverse nature and significance. The president advanced the following reasons for the crisis:

> The structural reasons are the change in the consumption pattern of energy in the industrialized countries; the substitution of oil by coal; the conservationist measures taken in, and the discipline and saving capacity shown by those countries. Temporary reasons include high inventories, the economic recession and the contribution of new sources of oil such as Mexico and the North Sea.[16]

Among the measures taken by the government were:

Cutting public expenditure for 1982 by Bs 8.4 billion, about $2 billion and reducing the budget of all public corporations by 10 percent

Increasing threefold the price of high-octane gasoline

Eliminating import subsidies

Substantially increasing the departure tax for all Venezuelans traveling abroad, from Bs 80 to Bs 300 per head

Conserving petroleum by various measures—an 80-kilometers/hour in all highways, special hours for service stations, the extension of the mandatory stop day for private vehicles in large cities of Venezuela, and the progressive use of liquefied petroleum gas in replacement of gasoline

Imposing the Buy Venezuelan decree and prohibiting imports of men's clothing and all types of shoes for a 1-year period

Increasing financial help to the agricultural sector

In addition, President Herrera said that all major projects in the electrical, steel, shipyard, and mineral fields would continue. He also said that the food-stamp program would not be canceled and that the acquisition of F-16 jet fighters would proceed as planned. Finally, he vowed that the bolívar would not be devalued.[17]

Some of these measures were definitely a step in the right direction; others not. Increasing the price of gasoline would give government an additional $1 billion per year, but the failure to increase the price of diesel oils, also a valuable and underpriced product in the domestic market, suggested to many Venezuelans that the gasoline-price increase did not respond to a coherent conservationist strategy but simply to a short-term government requirement for more funds. The food-stamp program and the buying of war planes, coming at a time in which extreme austerity should be exercised, threw serious doubt on the judgment of government leaders. Very expensive projects such as the steel complex for the State of Zulia and the shipyard for the State of Falcón also seemed to be totally out of line with the country's very tight economic situation.

Not surprisingly, Acción Democrática issued a document stating that the government was "leading the country toward bankruptcy." They mentioned "administrative paralysis, absence of imagination, improvising and contradictory measures" as characteristics of the government's performance during the years 1979–1982.[18] What seemed clear was that the measures announced by President Herrera completely failed to inspire the desired confidence and capital flight continued unabated at the rate of some $15-25 million per day. The measures failed to inspire confidence essentially because they were not coherent and because the government did not appear committed to them. Only a few weeks after announcing the gasoline-price increase, President Herrera confessed, "I was strongly opposed to an increase in gasoline prices, but the pressure from the opposition parties, Acción Democrática, the Workers' Central Union, Fedecámaras, and the media was too great."[19] This was hardly the type of announcement that could reassure the people about the quality and determination of the country's leadership.

The Crisis Deepens

During April OPEC named a committee made up of Venezuela, Algeria, Indonesia, and the United Arab Emirates to monitor the agreement on production quotas. Petróleos de Venezuela was now realizing the full impact on its earnings of the oil-price decline. They presented National Congress, April 22, with an estimate of Bs 25 billion, about $5.5 billion, as the most probable decline in oil income for the year. Minister Calderón Berti claimed, however, both before and after a trip to Vienna that "oil prices were improving and that Mexico was keeping exports below 1.25 million barrels per day for solidarity reasons with other oil-exporting countries."[20] In fact the price increase was restricted to the high-sulfur residuals, and Mexico was only exporting about 1 million barrels per day simply because that was all the country could sell.[21]

In numerous press releases, Minister Calderón Berti stated that "the inevitable evolution of the markets will tend to favor the demand and oil prices in the next months of this year."[22] Nobody took these statements as predictions anymore; rather, they were considered political announcements designed to tranquilize the public while the government waited for a new miracle: an unusually harsh winter in the north, a political upheaval in the Middle East. Capital flight continued at the alarming rate of over $100 million per day. Meanwhile, a study by the International Monetary Fund, "Perspectives of the World Economy," predicted that the world petroleum market would stay weak throughout 1983.[23]

In June 1982 Finance Minister Luis Ugueto contacted Acción Democrática and proposed to the party the partial utilization of PDVSA's oil-investment funds to pay for non-oil-related debts.[24] The party categorically refused to support this proposal and denounced it publicly. Minister Calderón Berti once more went to National Congress to ratify his optimism and to confirm his prediction about "the bright oil prospects for the second half of the year."[25]

Within Petróleos de Venezuela morale did not seem to be improving. A newspaper report observed:

Politics (has) appeared at the level of the (oil) operating companies. Appointments, promotions, salary increases, changes, and decisions which should be based exclusively technically (were) now made according to party lines. Alfonzo Ravard (claim our sources) has opposed this vigorously, but he has not been able to do a successful job against the executive will . . . . In hushed tones, but not less anxiously there is talk of negotiations and sinecures. The move of Meneven to Puerto La Cruz, it is said, is not independent from certain maneuvers in the real estate world.[26]

A very harsh confidential analysis advanced the following comments:

Ministerial interference has been evident in (managerial) decisions, which a now-pliable PDVSA board appears incapable of opposing effectively. TMR believes that many of those decisions are incorrect and that they will be harmful to PDVSA's future efficiency and profitability. . . . Decisions which should be made by PDVSA in accordance with the industry's criteria, experience, and management ability are being made by the Energy Ministry. Many experienced executives and technicians already have left the industry.[27]

The report contained some examples of the ministry's interference in PDVSA's operations such as arbitrary allocation by the ministry of production cuts to the different operating companies, without regard to existing commitments. This, according to the report, "hardly promotes Venezuela's image as a reliable source of supply." The decision to reassign existing operating areas to other companies without the benefit of an analysis of advantages, disadvantages, and costs involved was cited, as was the utilization of PDVSA's oil-reserve fund for 1982's PDVSA capital outlays. The analysis suggested that this money could in fact end up serving other purposes. Revisions of the Orinoco heavy-oil project to eliminate the required 300-megawatt electric plant were mentioned. The report considered this as illusory savings since the plant would have to be eventually built at a much higher cost. The president of Acción Democrática, Gonzalo Barrios, expressed fear that PDVSA's oil-investment fund could be utilized for purposes other than those for which it was created.[28] The party actually called for a national front to defend the integrity of the fund.

OPEC Meets in Vienna

In the first week of July OPEC met in Vienna amid considerable uncertainty. Spot market prices were closer to $31 per barrel than to the OPEC-agreed price of $34 per barrel. The oil market was, according to the *New York Times,* "moribund, almost dead."[29] Venezuela and other OPEC countries of high absorptive capacity were already feeling the financial squeeze. Minister Calderón Berti threatened OPEC member countries with an increase in Venezuelan oil production unless the other OPEC members respected the production quotas agreed upon in the previous meeting.[30] Iran was already producing 1 million barrels per day above its permissible level, Libya was 250,000 barrels per day in excess, and Nigeria about 300,000 barrels per day higher. Calderón Berti's warnings were not well timed. Venezuela was a member of OPEC's production-monitoring committee and should not have resorted to public threats to make a point. Furthermore the Venezuelan position had not been consulted with the other political sectors of the country. It has been suggested that the Venezuelan government had made the decision to increase production even before Minister Calderón Berti went to Vienna. The failure of other members to respect their quotas would have merely been the excuse for Venezuela to follow suit.

Not surprisingly, the OPEC meeting ended in almost total disarray, every country on its own. Returning to Venezuela Calderón Berti said, "I assume full responsibility for the increase in oil production."[31] Criticism of his unilateral action was immediate, however. Acción Democática opposed the production increase and blamed PDVSA's board for its "passive silent attitude in the presence of the irresponsible petroleum policy of the government."[32] The analysis made by TMR stated that

> The minister completely fouled up Venezuela's position at the July OPEC meeting. This disaster occurred in two fronts, each politically sensitive. The first error was to threaten publicly that Venezuela would ignore the OPEC production quota unless delinquent producers agreed to conform. . . . On the Venezuelan scene, the minister's conduct (was) even harder to understand. If Calderón had talked to Acción Democrática before (going to) Vienna, the party would have probably agreed to a production increase.[33]

As it was, TMR added, "They felt they had been deceived by the government. . . . Calderón Berti's judgment and competence must once again be called into question." Meanwhile the government was getting its hands ever closer to PDVSA's oil-investment fund. The president of the Venezuelan Central Bank stated: "Circumstances might suggest the employment of the oil-industry financial resources for other rentable project or to use the (oil) industry as a means of obtaining international funds for these projects."[34] Díaz Bruzuál was suggesting a road for PDVSA that had already proved disastrous for Mexico's PEMEX, which by 1982 had accumulated a foreign debt of $25 billion.

After the OPEC meeting ended in failure, spot oil market prices dropped further, to about $30 per barrel. Venezuelan production increased to

1.8 million barrels per day. Mexico increased exports by 250,000 barrels per day, again cutting prices of heavy oil by $4 per barrel below the Venezuelan average.

The End of the Oil Industry's Self-Financing

In August journalist C. Chavez reported that the government was seriously contemplating taking over PDVSA's oil-investment fund, a move he denounced, saying that such an event would seriously diminish the managerial autonomy of PDVSA.[35] In a speech celebrating PDVSA's seventh anniversary, R. Alfonzo Ravard stated,

> Thanks to its financial self-sufficiency Petróleos de Venezuela can overcome those obstacles derived from unfavorable international conditions. . . (self-sufficiency) must be maintained as the key to an uninterrupted financing of national development. . . .
>
> Moreover, self-sufficiency is indispensable to our freedom of judgment. If industry did not have it, it would have to ask the state for funds or borrow funds abroad. The first alternative would be contradictory, because the industry supplies most of the funds to the state. The second alternative would lead to a serious erosion of our freedom of decision.[36]

It is now evident that Alfonzo Ravard already knew that the government was seriously thinking of taking such a decision and was warning the country that, if this happened, it would be a tragic mistake.

Former oil minister Valentín Hernández made a public statement for the first time since moving out of public office:

> We have no doubts as to who should manage the oil industry. It cannot be the ministry. . . . If the country knew that the oil industry was being managed with little political interference and with clarity of objectives, there would not be a need for debate. However, this does not seem to be the case.[37]

By August 1982 the international reserves of Venezuela had decreased by more than $3 billion and capital flight showed no signs of decreasing. In these days Mexico nationalized its banking system while at the same time ordering a number of F-5 fighter planes from the United States.[38]

Venezuela was now the sixth overall supplier of oil to the United States, lagging badly behind Saudi Arabia, Nigeria, the North Sea, Mexico, and Canada and was beginning to be challenged for the U.S. residual-fuel-oil market by Mexico.

The first week of September the president of PDVSA went on television to say, once more: "The PDVSA oil investment fund should not be taken away from the industry," and the former President of Venezuela, Carlos A.

Pérez, denounced the intention of the government to do just that. He warned that such a decision "would mean the death of PDVSA." In spite of all warnings and opposition, the government authorized the Venezuelan Central Bank, September 28, to take over PDVSA's oil-investment fund. The decision was technically impeccable since, by law, all Venezuelan foreign reserves should be under the control of the Central Bank, but it represented a clear violation of the agreement reached by all political sectors at the time of the nationalization of the oil industry. It was then agreed that the industry would keep a fund exclusively dedicated to finance its capital projects so as to preserve fully its managerial autonomy. The stated objective of the present move, that of increasing Venezuela's power of negotiation vis-à-vis the international financial system, had probably been better served by leaving the oil-industry fund in the hands of its professional managers. As it happened, the international banking system had more reasons than ever to be wary of the Venezuelan economic situation.

Internally the decision of the government did not seem to restore public confidence. On the contrary, many Venezuelans saw it as one more step toward national economic disaster, as a totally counterproductive move. In a letter to President Herrera, Acción Democrática stated,

> It is not an exaggeration to think that (this action) could allow the Central Bank the use of the funds to satisfy other requirements. Such a possibility would militate against the financial autonomy of the industry and seriously damage its development, which is closely related to the opportune use of foreign exchange.[40]

Fedecámaras, through its vice-president A. Célis, protested that it was indispensable to preserve PDVSA's full financial self-sufficiency.[41] AGROPET, the organization of oil-industry technicians, declared that the decision of the government diminished the managerial autonomy of the industry.[42] It opened the door, they claimed, to the possible utilization of the oil-investment fund to cover temporary government deficits and could force PDVSA to borrow in the international finance markets. In the opinion of AGROPET, the move had been inspired almost solely in the government's short-term fiscal needs.

Minister Calderón Berti and President Herrera vigorously defended the petroleum policies of the government.[43] Although Venezuela was producing 2 million barrels per day, 0.5 million barrels above OPEC's originally agreed level, Herrera asserted that "Venezuela (continues) to produce oil based on (our) requirements and convenience and on the needs of (our) clients. . . . The country maintain its purpose of unity and discipline within OPEC." Curious words, indeed.

Upon his return from a trip to the Far East, PDVSA's President R. Alfonzo Ravard called a press conference to express surprise for the govern-

ment's decision and his hopes that the move would be "a temporary one."[44]
Next day, E. Sugar, former vice-president of Lagoven and an oil executive
with 35 years' experience, said that the government interference in the management of PDVSA "was already intolerable."[45]

Christmas 1982

Starting the first week of December, Venezuela cut the prices of distillates
by $3.36 per barrel. General R. Alfonzo went to Congress to say that
"PDVSA had lost jurisdiction over its funds."[46] He added that since 1976
the intention had been to give PDVSA managerial autonomy and self-
financing and that such an intention had been explicitly written into the bylaws of the corporation. Acción Democrática, URD, and MAS seemed
determined to modify the government's decision by voting a modification
of the degree issued by government so that it would not apply to PDVSA.

Winter in the United States and Europe appeared to be mild. As a result
the oil market continued to be extremely weak. West Germany's oil
imports, for example, were 9 percent lower than in the same month of 1981.
Faced with an increasingly critical situation, the Venezuelan finance minister resigned. The government asked PDVSA to utilize about Bs 7.5 billion,
some $1.8 billion, to bail out the Workers' Bank, the largest Venezuelan
bank, which was bankrupt. This bank had been the object of systematically
corrupt mismanagement and now required an urgent injection of fresh
money. When the already highly weakened and demoralized board of
PDVSA refused to accept this decision, Minister Calderón Berti called a
special shareholders' meeting and ordered the board to acquire up to $1.8
billion in public debt bonds. The board of PDVSA meekly agreed.[47]

Meanwhile OPEC was getting ready to meet again. Minister Calderón
Berti had decided to visit Kuwait, Iran, Algeria, and all the oil capitals, in
what was hailed as a very intense and personal diplomatic effort, to try to
obtain an agreement for an individual-country production allotment. When
the OPEC meeting ended, however, there was no agreement on production,
no visible progress. Although OPEC had decided to keep the marker crude
price at the level of $34 per barrel and overall production had been
increased to 18.5 million barrels per day, individual member countries were
not committed to any production target.

Without external relief in sight, the Venezuelan government now turned
full attention to PDVSA's reserve fund. The Central Bank ratified the utilization of up to $1.8 billion from the fund in the acquisition of public debt
bonds. An editorial of *El Diario de Caracas* commented on this decision as
follows:

In a short 3 months the government took away from the oil industry the control of the funds it generated and, in spite of all warnings to the contrary, it also took Bs 7.5 billion which were earmarked for oil-related projects to (bail out) the Workers' Bank and to cover the deficits in the (national) budget.

As a result, all the development programs of the Orinoco heavy-oil area are being frozen until it is known how much can be executed in an organic manner. . . . There is nothing now to guarantee PDVSA that the government will not demand, tomorrow, another Bs 7 billion loan, backed by more public debt bonds so "solid" and "desirable" that nobody would accept them even as a gift. . . . Within the oil industry, in the minds of tens of thousands of employees, the fundamental doubts now are: What will happen tomorrow?, What should we be ready for?, How can we keep the industry going? . . . In one single act they (the government) destroyed the political agreement, the national agreement which protected the oil industry from the fate of other state-owned enterprises; they forgot the letter and spirit of (those) agreements.[48]

The confusion, now almost total, was utilized by the political extreme left to demand more control of the oil industry. The secretary general of MAS, P. Márquez, wrote, "We are glad to see that the (Orinoco) programs are being adjusted downwards. . . . We insist on the rigorous control of these programs by Congress."[49] Thus the vicious circle of politicization, industry deterioration, and more politicization was completed. Christmas 1982 had brought not only deep economic crisis but also disenchantment, gloom, and spiritual depression to the Venezuelan oil industry and, worse, to the country.

The Nature of the Crisis

In 1982 the attacks on the Venezuelan oil industry were no longer disguised or muted. The president of the Central Bank, L. Díaz Bruzuál, had already said in December 1981 that the oil industry had "very low productivity," and that "When employment increases in 50 percent . . . and salaries are doubled, there is no doubt that the industry is being poorly managed."[50] The president of PDVSA replied to this that "The superficial analysis of cost and manpower growth as compared to a basically stable production leads to wrong conclusions."[51]

The president of PDVSA was essentially right. To produce a barrel of Venezuelan oil was, inevitably, more expensive in 1982 than in 1976. Díaz Bruzuál was essentially wrong, but he had the firm belief that he was right. One year later this sincere conviction would lead him and the government to endanger gravely the industry's capacity for self-financing.

The real problem, as lucidly suggested by J. Olavarría in a *Resumen* editorial, was the extreme dependence of public expenditure on oil income, the fact that government has always spent up to the last cent derived from oil and had established a spending pattern which now seemed difficult to modify.[52] At least since the 1960s, the Venezuelan government had been spending all the money received from oil. Even in 1974 when oil revenues suddenly tripled, expenditures tripled as well. Now the drop in oil income was seen by the Venezuelan political sector as the fault of the oil industry. In fact the fluctuations in world oil demand were essentially beyond the control of the industry. What was within the power of the country to modify was the level of public expenditure.

The 1982 budget, however, was structured without a proper analysis of the effect that the world oil market situation could have on the level of Venezuelan oil income. The estimate of oil exports for 1982 was of 1.8 million barrels per day, at an average price of $30.45 per barrel. But by March 1982, Venezuela was producing 1.5 million barrels per day and exporting only about 1.3 million barrels per day at an average price of some $29 per barrel. It was evident that the gap between prediction and reality was enormous and required dramatic adjustments—which were not forthcoming. Minister Calderón Berti kept calling press conferences to say that there would be an increase in the oil demand "in the second semester of 1982," a prediction for which there was no basis whatsoever. President Herrera kept talking of grandiose projects such as a bridge to connect the island of Margarita with the mainland and costly programs such as a food-stamp program, which would cost more than Bs 3 billion, a sum that the country simply could not afford.[53] As a consequence the public debt of Venezuela kept creeping up, reaching about $20 billion by mid-1982, and uncontrolled capital flight cut deeply into the country's international reserves. The financial crisis had now become a crisis of confidence and the crisis of confidence threatened to become a political crisis.

The danger of a political crisis was obvious, since the country was already in the midst of a presidential electoral campaign. This complication minimized the possibilities of a broad political pact to soften the negative impact of the financial crisis. Contrary to previous national crises, in which the majority parties had successfully resorted to the concept of political or parliamentary pacts and coalitions, this time political conditions were probably the worst since the country had returned to democracy. The political parties were deeply torn by internal strife, and the quality of the political exchange had reached an all-time low. The government of President Herrera was almost totally lacking in popular support and, worse, was incapable of seeing with clarity the true extent of the dangers being faced by the country.

Against this gloomy background it was reasonably easy to predict what might happen. If a broad political pact to overcome the financial difficulties was not probable, the government would simply continue reacting to external stimuli rather than taking steps to prevent further damage. Its main concern would tend to be the short-lived solution, the patch-up approach, rather than the search for stable, long-term answers.

The example of Mexico did not seem to help Venezuela. In 1976 Mexico had abandoned all pretenses of austerity and had embarked, under President López Portillo, on "a series of grandiose projects," including the construction of twenty nuclear reactors.[54] The financing of these projects required extensive borrowing abroad. By 1982 Mexico's debt was the largest in the world, up to $80 billion, and the oil glut had severely diminished the value of Mexico's oil exports. Yet the Mexican government continued to borrow money as "if nothing had changed. . . . as a result Mexicans began to lose confidence in their currency, rushing to buy dollars and thus undermining the peso. But López Portillo would not devalue the peso . . . until it was too late."[55] The development of the Venezuelan financial crisis closely paralleled the Mexican situation, but considerable effort was made by Venezuelan politicians to emphasize the differences between the two cases. In spite of all warnings, Venezuela continued to march inevitably toward disaster.[56] As in the case of Mexico, Venezuela had allowed the fate of the country to depend entirely on the petroleum market and had not made any provisions for the collapse of that market.

It was therefore not surprising to see that, faced with an acute shortage of funds as the end result of lower oil income and unwise levels of public expenditure, the government decided to take over PDVSA's oil-investment fund. This was the type of short-term solution that would be made by a government without many alternatives left. When part of the fund was further earmarked by the government to bail out the corrupt and bankrupt Workers' Bank, everybody in the country saw this as one more predictable step in a choreography of doom.

The prevailing political atmosphere was further complicated by the fact that dissent was being taken by the government as disloyalty. Maraven's president warned publicly against this aberration of democracy when he said,

> When we confuse disagreement with disloyalty we are contributing to the creation of a society of eunuchs. This is particularly dangerous in Venezuela, where the state owns an enormous percentage of the means of production and services: oil, petrochemicals, iron, aluminum, electricity, telephones, food. Those managers who, in the view of the state are disloyal because they dare to indicate disagreement might find their progress blocked. The great power of the Venezuelan state, as opposed to the power of the state in those countries where there exists a strong and diversified public sector, can make success impossible to attain for this group. This is a

very dangerous and totalitarian situation. In the last years we have witnessed too many changes in the professional management of state industries not related to the competence of those who got out or in but, rather, based on the existence of disagreement (wrongly) defined as disloyalty.[57]

The Short-Term Outlook

The short-term outlook is bleak. Even assuming complete rollover of the external debt, the official reserves of Venezuela, which were almost $20 billion in 1981, might well drop to about $9 billion in 1983. The total national debt could move up to about $30 billion, while the portion maturing in under 1 year could be as large as $14 billion.

Oil export prices, which averaged about $29.50 per barrel in 1981, are closer to an average of $25 per barrel in 1983. Internal consumption will be about 400,000 barrels per day and this will liberate only an average of 1.5 million barrels per day for exports. This could give PDVSA a total income of about $13 billion for its oil exports. Even assuming that industry costs remain essentially constant, the income before income tax would be about $9–10 billion or about $5 billion less than in 1981. Fiscal income derived from oil would, therefore, be approximately $8–9 billion, about 60 percent of what it was in 1981. Will the country be able to adjust its budget to this new level of oil income, which seems likely to remain unchanged at least throughout 1983?

If the answer is no, then massive external borrowing will inevitably continue. The answer might well be negative, given the fact that Venezuela will be in an electoral year rendering the political environment far from favorable to the attainment of a broad political agreement directed to solve national economic problems. But there seems to be very little doubt that without such a broad political agreement, hope for a recovery of the Venezuelan economy can be slight at best.

In August 1983 a new board of Petróleos de Venezuela will be selected. It is imperative that this selection conform to the rules of promotion by merit and apoliticism which had essentially characterized oil-industry appointments until 1981. If the tendency toward political appointees or cronyism is reinforced in August, the moral and financial deterioration of the Venezuelan oil industry will probably pass beyond the point of no return.

Notes

1. G. Coronel, "Oil, Towards an Age of Balance," University of Tulsa Annual Report 1981–1982, Tulsa, 1982.

2. *El Nacional,* February 17, 1982, p. A-22.

3. *El Diario de Caracas,* March 26, 1982, p. 37.

4. *El Universal,* February 19, 1982, p. 2–32.

5. *El Diario de Caracas,* February 19, 1982, p. 32.

6. Alvaro Silva Calderón, MEP's oil expert, quoted by *El Universal,* February 24, 1982, p. 1–12.

7. The quote is from *El Nacional,* February 25, 1982, p. D-1.

8. *El Universal,* March 1, 1982, p. 2–9.

9. *El Universal,* March 1, 1982, p. 3–7.

10. *El Universal,* March 3, 1982, p. 1–13.

11. *El Diario de Caracas,* March 3, 1982, p.6.

12. *El Diario de Caracas,* March 29, 1982, p. 2., and *El Universal,* March 29, 1982, p. 2–1.

13. *El Universal,* April 2, 1982, p. 2–1.

14. *El Universal,* April 2, 1982, p. 1–30. A similar thought was expressed by J. Olavarría in *Resumen* 436, p. 3.

15. *El Universal,* April 2, 1982, p. 2–4.

16. *El Universal,* April 10, 1982, p. 2–1.

17. Ibid.

18. *El Universal,* March 10, 1982, p. 1–1.

19. *El Universal,* April 15, 1982, p. 1–13.

20. *El Universal,* April 16, 1982, p. 1–17, and *El Nacional,* April 27, 1982, p. D-8.

21. *Petroleum Intelligence Weekly,* quoted by *El Nacional,* May 7, 1982, p. A-4.

22. *El Nacional,* April 28, 1982, p. D-13; *El Universal,* May 17, 1982, p. 2–13; *El Universal,* May 3, 1982, p. 1–14; *El Universal,* May 27, 1982, p. 1–17.

23. Cited in El Universal, May 30, 1982, p. 2–17.

24. *El Nacional,* June 15, 1982, p. D-11.

25. *El Universal,* June 21, 1982, p. 2–2.

26. *El Diario de Caracas,* June 20, 1982, p. 5.

27. *The Monthly Report,* June 27, 1982.

28. *El Nacional,* June 25, 1982, p. D-1.

29. Quoted in *El Nacional,* July 7, 1982, p. D-1.

30. *El Universal,* July δ, 1982, p. 1–1.

31. *El Universal,* July 26, p. 1–13.

32. *El Universal,* July 28, 1982, p. 1–12.

33. *The Monthly Report,* July 25, 1982.

34. *El Universal,* July 24, 1982, p. 1–12.

35. *El Universal,* August 21, 1982, p. 1–16.

36. *El Universal,* August 29, 1982, p. 2–1.

37. *Número,* August 23, 1982, p. 14ff.

38. *El Diario de Caracas,* September 2, 1982, p. 39.

39. *El Nacional,* September 10, 1982, p. D–17.

40. *El Universal,* September 29, 1982, p. 1–12.

41. *El Universal,* September 29, 1982, p. 2–1.

42. *El Nacional,* October 7, 1982.

43. *Ultimas Noticias,* October 23, 1982, p. 32, and *El Nacional,* October 18, 1982, p. A–1.

44. *El Nacional,* November 3, 1982, p. D–1.

45. *El Diario de Caracas,* November 3, 1982, p. 24.

46. *El Universal,* December 2, 1982, p. 2–1.

47. *Resumen* 473, December 26, 1982.

48. *El Diario de Caracas,* December 15, 1982, p. 6.

49. *Ultimas Noticias,* December 15, 1982, p. 104.

50. *Resumen* 426, January 3, 1982, p. 17.

51. Ibid.

52. *Resumen* 436, March 14, 1982, p. 3.

53. For an analysis of the myopic attitude of the Venezuelan government during the crisis, see *Resumen* 440, April 11, 1982.

54. See *Time,* December 20, 1982, p. 39ff. Reprinted by permission, 1982, Time Inc.

55. Ibid.

56. In July 1982 three of the most important government bureaucrats in the country: The president of the Central Bank, the president of PDVSA, and the general comptroller, warned publicly that Venezuela was heading towards economic disaster (see *Resumen* 462).

57. *Resumen* 471, November 14, 1982, p. 7.

**Part III
Outlook**

13 National Oil Companies

The number of national oil companies has grown consistently throughout the years. With notable exceptions such as the United States, most consuming and producing countries have a hydrocarbons agency, usually an integrated company. Argentina created a petroleum bureau as early as 1910 and an oil company, Yacimientos Petroliferos Fiscales, in 1922. The creation of Compagnie Française de Petrole (CFP) in France and AGIP Italy also date back to the early 1920s. Some other important national oil companies such as the British National Oil and Petro-Canada, like Petróleos de Venezuela, are of more recent creation, in the 1970s.

These companies have been formed under different cultural, political, and economic environments, in countries with different amounts of hydrocarbon resources. Many of the reasons to create them have been very similar, however. L.G. Grayson lists the following group of objectives guiding the creation of national oil companies in industrialized countries: to become more knowledgeable about the hydrocarbons industry; to reduce national dependence on foreign firms; to secure volumes of "inexpensive" oil; to increase national technical expertise in the field of hydrocarbons; to utilize them as political or development tools; to deal with OPEC.[1]

Many of these objectives could also apply to national oil companies in developing countries. But in addition national oil companies in the developing countries often incorporate other aims. They are created to serve as catalysts of national economic growth, as tools for social improvement, as a means to exercise control over a basic industry. Behind the creation of most national oil companies there is, more often than not, an ideological motive, be it in England, in India, or in Venezuela. Usually ideology, as in England, is part of the philosophy of government but the most common force is national pride.

H. Madelin sees the creation of state oil companies as the result of a progressive change in the relationship between states and the oil industry. He distinguishes three stages in this relationship: (1) the *liberal* model, in which petroleum deposits go to the high bidder and legislation corrects tendencies toward capital concentration; (2) the *general regulation* model, in which government decides on a policy or policies and issues laws, decrees, and instructions within which companies have to work; and (3) the *direct management* model, in which government takes over property and sets its own organization. Not all national oil companies have gone through this long and systematic evolution. Petróleos de Venezuela is a classic product of this evolution but Petróleos Mexicanos, for example, never went through the generally long-lived regulatory stage. The Argentinian national oil company did not follow this process at all, being the product of an early political decision based on a deeply nationalistic philosophy. The same can be said of the events leading to the creation of Petrobrás in Brazil.

National oil companies can be said to belong to two main groups: those from net oil-consuming countries and those from oil-producing and oil-exporting countries. The companies from the first group generally exist to acquire, at the lowest possible cost, the hydrocarbons required by their shareholders. The companies of the second group aim at selling, at the highest possible price, the hydrocarbons that they produce and export.

National Oil Companies that Buy

This group essentially comes from industrialized countries, although they also exist in Third World countries such as Chile, Uruguay, and India, and are net oil importers. These companies have usually been created as the result of a deliberate political strategy, to assert national control over the use of energy and to regulate the domestic market of hydrocarbons in their respective countries. Another major objective is to represent their governments in dealings and negotiations with national oil companies from producing countries. Examples of this groups are the Compagnie Française des Petroles (CFP), the Ente Nationale Idrocarburi (ENI), and Petróleos Brasileiros (Petrobrás).

CFP

This is the largest national oil company in Europe, selling about $9 billion dollars per year.[3] It was established in 1924 after the First World War had shown the vital importance of oil during an armed conflict. CFP was not a solely state-owned enterprise, however. Multinational oil companies and private French companies were allowed to own shares, but the control was in the hands of the French government. The original objective was for CFP "to follow specifically French interests as Anglo-Persian followed British interests."[4]

During the early stages of its activity, CFP primarily engaged in exploration, from Venezuela to the Middle East. In 1929 a refining subsidiary was created, Compagnie Française de Raffinage (CFR) to refine up to 25 percent of French domestic requirements. During the 1930s and 1940s CFP slowly grew into a large integrated oil company, and in 1946 it obtained "a majority holding in a distribution company in West Africa and thereby established its first distribution outlet outside France."

The president of CFP after the Second World War, Victor de Metz, "wanted to run CFP like a multinational oil company."[5] Due to his leadership, the company considerably expanded its activities. By 1957 it had subsidiaries in Qatar, Algeria, Portugal, Yugoslavia, Iran, Italy, Australia, Abu Dhabi, and several other countries. Already by the 1960s CFP was referred to as the eighth sister and had entered the Canadian and the U.S. market. In the 1980s CFP has become a very large, integrated oil company, with very little diversification in petrochemicals or other energy sources.

CFP's president is named by the French government. The board is made up of representatives from the different shareholders, including companies such as Elf-Aquitaine. It comes under the supervision of the Ministry of Industry.

In general CFP has emphasized international activities rather than domestic activities. Its partially private ownership has combined with its international character to increase its level of managerial autonomy. Some of the other characteristics which have contributed to the relatively high level of managerial autonomy of CFP are the sophisticated technology typical of the oil industry, the highly competitive atmosphere in which the company operated, and the strength of its top executive officers.[6] CFP has been reasonably successful financially. According to Grayson the profitability of the French oil company (defined as net income over shareholders' equity times 100) has been comparable to that of companies such as British Petroleum, although in the last 20 years it has been much lower than the percentages for companies such as Shell, Phillips, and Petrofina.[7]

ENI

The Italian national oil company was established in 1953 as a holding company for all companies operating in the petroleum industry. It included the original national oil company, AGIP, created in 1926, as well as four refining companies, a natural gas company, a marketing company, and ANIC, a chemical organization. From its creation ENI was controlled by Enrico Mattei, a vigorous personality who ruled the company with an iron hand.

The chairman of ENI is chosen by the government. Throughout the years the nature of the appointment has changed from that of strong, exceptional managers such as Mattei to that of purely political appointees, such as the naming, in 1979, of Socialist Giorgio Mazzanti in a move that also sent a Christian Democrat chairman to IRI (Instituto per la Ricostruzione Industriale) and a Social Democrat chairman to EFIM (Ente Partecipazione e Finanziamiento Industria Manifatturiera). The mechanism of selecting the management team also evolved with time:

> Enrico Mattei, ENI's founder, was a powerful man with clear ideas as to how the company was to be managed. His original staff was composed of collaborators. . . . In all cases, professional capability was a prerequisite for filling any position, regardless of its level. Today, ENI's hierarchy reveals other influences. One of the six major subsidiaries owned by ENI is headed by a political appointee who is clearly the puppet of higher management.[8]

Mattei was a strong factor in ENI's rapid expansion during the 1950s and 1960s. The company was soon subject to two deforming pressures. One was the Mattei's own ambition, which drove him to buy a newspaper (*Il Giorno* of Milan), a news service, and an advertising agency.[9] The other was the government's request that ENI invest in the economically depressed areas of southern Italy.

"Wherever oil and natural gas have been discovered, ENI has been asked to provide jobs for the people of the area."[10] The investments forced upon ENI included textile and clothing companies purchased

> at the government's request. . . . In one important case the government made it clear there would be an increase in the tax on natural gas unless ENI took over a struggling textile company in the Alps and two clothing plants, one on the Adriatic Sea and the other south of Rome. In another case, the Communist party promised not to filibuster against an increase in ENI's endowment fund provided ENI rescue a hat-manufacturing concern in a red town in Tuscany.[11]

For many years ENI kept very close ties with the Christian Democratic party. In the 1976 elections, however, the Christian Democratic party lost ground to the Communist party and had to share with them much of their power, presumably including that over ENI. In 1979 the chairmanship of ENI went to a member of the Socialist party. His appointment lasted less than a year since he was mentioned in a case of commissions being paid to ENI officials as part of a deal between Italy and Saudi Arabia. His replacement resigned "one day before he was scheduled to assume the position . . . alledgedly because of increasing governmental interference into ENI's affairs. Thus, the chaotic Italian political situation finally was transferred, in full, onto ENI."[12] By 1975 ENI had accumulated losses of more than $4 billion. A comparison between ENI and some private oil companies for this year showed the results listed in table 13-1. These figures illustrate the main differences between profit-oriented enterprises and a politicized state-owned organization. ENI had from 2.5 to 3 times more employees and less production than Shell or Standard Oil and, in spite of having a similar level

Table 13–1
Financial Results, ENI and Other Companies

	Standard Oil, Indiana	Shell Oil, United States	Phillips Petroleum	ENI
Gross income ($ millions)	10,025	8,418	5,106	8,748
Net income ($ millions)	970	620	430	(89)
Employees	42,217	32,287	30,802	92,420
Production (million tons/year)	45.6	30.6	11.2	16.7

of gross income to that of Shell, it yielded a loss rather than a profit. Even when allowing for different production and marketing patterns, it is hard to explain these performances by purely financial or operational reasons. The negative influence of politicization seems clear.

Petrobrás

Petróleos Brasileros, Petrobrás, was created in 1954 inheriting the assets of the Conselho Nacional do Petróleo (CNP). Its main objective was to find "enough oil to make Brazil self-sufficient."[14] The same year, the well-known American geologist Walter Link was hired as exploration manager in charge of organizing the department and planning the oil search. The report on the hydrocarbons prospects of Brazil prepared by Link and his technical team, including six top-level Brazilian geologists, concluded that the expectations for hydrocarbons in the country were modest and restricted to the smaller Bahia and Recife basins. The prospects of the Amazon basin were severely downgraded in the report.

This report had a great impact on nationalistic Brazilians, who believed that their country had enormous oil reserves. Its publication "hastened the politicization of Petrobrás, although this was perhaps inevitable in the face of changes that were taking place in national politics at that time."[15] The president of Petrobrás went to the Brazilian Congress to attack Walter Link. He considered the report a maneuver of foreign capital to take over the Brazilian oil riches. Gabriel Passos, the new oil minister under President Joao Goulart, declared that Petrobrás was under attack from the outside, "Linkism," and from the inside in the form of cronyism and political patronage. He fired the president of Petrobrás, G.C. Barroso, who had reached that position with the support of the Bahia oil workers' union. The workers battled against the new president of Petrobrás and, again, used their power to have him replaced by a man more sympathetic to their views, Francisco Mangabeira. In turn he assigned labor a seat on the board. This prompted the president of the Federation of Industries of Brazil to remark that "to be from Bahia and a nationalist should not be the sole criteria to become president of Petrobrás."[16]

As early as 1960 the leading technicians of Petrobrás publicly warned President Janio Quadros about the increasing politicization of Petrobrás. They claimed that there was an exodus of qualified personnel because of this.

Also in the early 1960s Petrobrás became a motor for the industrialization of the country. Under the motto: "The Oil Is Ours," Petrobrás began the construction of a refinery with local materials. The company also placed an order for six tankers from Brazilian shipyards and built a town in Nova Olinda. The Petrobrás unions "were able to gain an increasing control of the organization by aligning themselves with President Goulart and, even more so, with the radical left outside Petrobrás. The technical and administrative staff resisted this process, but there was a defensive struggle."[17] Slowly, however, Petrobrás technocracy became stronger. By 1974 "most senior positions, up to and including the board of directors, were filled by specialists from within the organization. . . . However, the company (was) relatively ill equipped to deal with a political environment which opened up, at first gradually and then, quite rapidly after 1974."[18]

Petróleos de Venezuela's own development, several years later, closely resembles the evolution of the Brazilian national company. Petrobrás had few friends. The private sector distrusted them. The labor unions and the leftist political parties did not like their technocrats, perceiving them as an arrogant elite. The press resented what they termed the "secrecy" of those technocrats. How very similar indeed to the perception that these very same sectors developed, in Venezuela, about their national oil company!

The politicization of Petrobrás, according to P.S. Smith, was due to several factors:

1. The inability of the company to become independent of government revenues
2. The myth about Brazil's oil wealth, which was used as an excuse to spend millions in a futile search
3. The exaggerated nationalistic environment in which the company operated
4. The influence of labor unions
5. The participation of the political sector in company decisions
6. The rapid turnover of executives
7. The fear of multinationals, which converted Petrobrás into an instrument of resistance against economic imperialism[19]

National Oil Companies That Sell

The second group of companies generally exists in oil-producing and -exporting countries such as the OPEC member countries and Mexico. They have been created to minimize the influence of multinational oil companies in the producing countries and to gain control of a vitally important world commodity. These companies usually are profit-oriented. Their tasks include the selling—at optimum prices—of hydrocarbons in the world markets. Examples are Petróleos Mexicanos (PEMEX), and Pertamina, the Indonesian National Oil Company.

PEMEX

From the early 1930s Mexico followed a policy of increasing government control of its oil industry. In 1934 the Cárdenas government created Petromex, a state oil agency to develop state-owned oil deposits with a view to supply the local market. President Cárdenas preferred to exert control in a gradual fashion. Vincente Lombardo Toledano, the leader of the strong Confederation of Mexican Workers, favored a more radical solution, that of nationalization. The opportunity came in 1937, when the oil workers asked for salary increases totaling more than 60 million pesos and the oil companies offered less than 15 million pesos. A commission of three experts, led by Jesus Silva Herzog, established that the equitable amount to be paid should be 26 million pesos. The companies refused. After some months trying to reach a compromise, Cárdenas expropriated the oil companies. G. Philip remarks that "it is difficult to escape the conclusion that, whereas in 1936 and 1937 it was the government which had been pressurizing the companies, in late 1937 and early 1938 it was the companies which took the more aggressive line.[20] Their publicity within Mexico suggested a desire to polarize the issue and face down the Mexican government, which, it was widely believed, could not afford the economic consequences of an expropriation and would in any case be unable to work the expropriated properties." Philip adds that "the Cárdenas government chose to expropriate not because it preferred confrontation to bargaining but because bargaining did not seem available."[21] Philip also makes the observation that "while the oil nationalization appears to have been a popular act, public opinion played a role which was essentially subordinate."[22]

Petróleos Mexicanos was created in June 1938. The oil workers wanted the company to be turned over to them for management. The final organization chosen by the government, however, was that of a board with six members named by the president and five members named by the workers' union. The director general and the top managers were also appointed by the president of Mexico. The government established the following policy regarding the organization of the nationalized enterprise:

1. The industry was restructured on a nationwide basis, eliminating duplication of functions carried out by the expropriated advertising, sales, and other firms.
2. The oil workers' union was reorganized to reflect the new structure of the industry.
3. Conferred upon PEMEX was the right to appoint "trusted" personnel to key positions.
4. PEMEX was authorized to eliminate positions it deemed unnecessary.
5. No pay increases were to be granted.
6. Merit was to be emphasized in making promotions.
7. Greater efficiency and output were to be demanded from workers.
8. Management was to be allowed a free hand in reassigning workers to other parts of the country.[23]

In spite of the government's intentions, PEMEX's budget grew by more than 26 million pesos in the 22 months following its creation (the amount which had originally caused the dispute); the reasons were "declining efficiency, (labor) irresponsibility, lack of discipline, and wholesale thefts."[24] Plants were closed down by workers in disagreement with managerial decisions. When the director general called on the workers to modify their attitude, union leaders asked for his dismissal.[25]

From the beginning PEMEX had social objectives which seemed to overshadow profit objectives. According to J.R. Powell, in his book on *The Mexican Petroleum Industry 1938–1950,* quoted by G.W. Grayson:

> Prices (of petroleum products) were based largely on social and political considerations. . . . It was widely felt, especially by labor, that PEMEX was not a business association required to make ends meet. . . . Management . . . seems to have developed an attitude of indifference since both prices and wages were set . . . by government policy which was subject to political influence.[26]

PEMEX's board of directors is chaired by the minister of National Patrimony and Industrial Development (SEPAFIN). The other five board members named by government include the minister of Finance, the minister of Commerce, the director general of Nacional Financiera, and two more high bureaucratic officers. PEMEX comes under the general coordination of a Ministry of Programming and Budgeting, a national planning group. The director general, since the times of Antonio Bermúdez, has been a powerful figure, very close to the president of Mexico. Bermúdez enjoyed an intimate relationship with President Miguel Alemán, first, and later with President Adolfo Ruíz Cortines. During his 12 years as head of PEMEX, Bermúdez resorted to heavy borrowing in order to make the company grow. From 1952 to 1958 the debt of PEMEX increased from U.S. $76 million to $311 million, still a very modest figure compared to the approximately $26,000 million, which represents today's debt. During 1958–1964 the López Mateos regime, there was considerable growth in the Petrochemical field. There were also growing signs of mismanagement, including the selling of jobs, the creation of private companies by PEMEX managers to contract with the state oil company, and the creation of union leader "baronies" such as the one enjoyed by Jaime Merino in Poza Rica. "Even López Mateos's personal pilot established a drilling company that did business with PEMEX."[27]

By 1958 PEMEX already was near bankruptcy. About 70 percent of its capital budget came from short-term loans. Jesús Reyes Heroles, the director general from 1964–1970, tried to curb the influence of the unions but allowed political appointees to obtain key managerial jobs.

Between 1964 and 1976, the work force of PEMEX moved from 50,000 employees and a payroll of 1.8 billion pesos to 95,000 employees and a payrol of 11.3 billion pesos. During the tenure of Antonio Dovalí Jaime, 1970–1976, the technical staff was forced to join the union. According to Grayson the technical staff were prepared to resign with Dovalí Jaime if he had stood firm, but he caved in to the demands of the union.[28] The only top executive to resign in protest against this decision was Francisco Inguanzo, the production technical head.

Jorge Díaz Serrano replaced Dovalí Jaime and became a very strong director general because of his very close friendship with President López Portillo. A period of great expansion took place, fueled by important new oil finds. During the López Portillo administration, the economic events of Mexico and the growth of PEMEX became even more intimately connected. The country tried to follow a very ambitious program of industrial and agricultural development based on the significantly increased oil revenues, but the collapse of the oil market in 1981 and 1982 created an economic crisis of major proportions. As Carlos Andrés Pérez in Venezuela, López Portillo tried to use increased oil revenues as a tool to convert Mexico into an instant industrial and agricultural giant. He said, "Oil is the pivotal point of our current economic policy and the means to obtain financial independence."[29] By the end of his administration, though, the country was in deep economic crisis.

During its evolution PEMEX developed certain characteristics and has been subject to constraints which have greatly influenced its overall efficiency:

It is perceived by the people as a symbol of economic independence and Mexican technical proficiency.

It is seen by government as a tool for national economic development.

Its director general usually maintains a very close personal relationship with the president.

The control of its board is shared by the government and the oil workers' union.

Financing has mostly been based on foreign, short-term loans. Because of this, PEMEX's foreign debt makes up about 40 percent of Mexico's total debt.

The organization has always had more employees than required, due to the pressure of the oil workers' union.

Jobs are "sold" by the union, not assigned directly by management.

The workers' union receive 2 percent of the value of contracts agreed by PEMEX with private firms and 50 percent of all contracts are granted directly by the union.[30]

Vacancies are commonly filled by the relatives of the employees who are contractually empowered to nominate them.

Domestic hydrocarbon prices are regulated by government and this penalizes significantly the financial performance of PEMEX. Gasoline sells for about half the price of the world market price, diesel oil for about one-fourth, and fuel oils for one-tenth of the world market prices.

PEMEX has mostly been a "technically talented but administratively chaotic institution which badly needs strong . . . leadership."[31]

The early PEMEX managers were technocrats with an intense loyalty to the organization. In time they have been replaced by administrators integrated into the Mexican political system.[32]

Pertamina

The national oil company of Indonesia was structured in 1968 to combine the activities of the existing state oil companies PN Permina and PN Pertamin. A third oil company, PN Permigan, had been eliminated in 1966 because of its connections with the Indonesian Communist party. Originally

Pertamina came under the control of the Ministry of Mines. In 1971, however, the law was changed to make Pertamina accountable to a national board of directors made up of the ministers of mines, finance, and planning. The law of 1971 attempted to take away from Pertamina much of its previous financial autonomy. The national board dictated the general policy for Pertamina, supervised its management, named the members of the executive board, approved the budget, loans, sales, and purchase agreements and the banks at which the funds of the corporation were deposited.[33]

Although state control over Pertamina was, at least on paper, extremely severe, this company soon became a classic example of mismanagement. President Ibnu Sutówo embarked on a diversification spree which took the company into petrochemicals (1971–1978), liquified natural gas (1973), an aviation company (PT Pelita), a hotel chain, a 300-bed hospital, 50 percent of an insurance company, 60 percent of a steel company (Krakatau), 50 percent of a fertilizer company, even a restaurant in New York City. By 1974 the budget of Pertamina represented about 40 percent of the Indonesian national budget. In 1973 Pertamina produced about 100,000 barrels per day and had some 32,000 workers whereas Caltex, operating in Sumatra, produced 1 million barrels per day and had about 4,000 employees. According to Fabrikant "numerous important military and political figures have long been economically (financially) dependent on Pertamina."[34] In spite of its inefficiency, the company became, because of its size, the key economic and political organism in the country. It seems evident that, in time, the company started to drift away from national priorities and became much more interested in advancing its own priorities. It utilized oil revenues for its own commercial diversification, and Dr. Ibnu Sutówo seemed to use the presidency of the company to gain considerable personal power. The example of Pertamina clearly seems to suggest that formal controls are not as important in determining success of a state enterprise as are the personal honesty and competence of its managers.

Common Patterns of Evolution

The highly condensed examples of state oil enterprises given above, two from developed countries and three from developing countries, seem to suggest a similar pattern of evolution applicable to most companies, regardless of their national origin. The differences seem to derive more from the nature of the ownership of the enterprise, whether only partially state owned like CFP or fully owned like the rest or from the excellence and honesty of management than from any other characteristic.

The Conflict between Social Objectives and the Profit Motive

Many state oil companies seem to pursue social objectives in preference to a basic profit objective. This is true of Petrobras, Pemex, and Pertamina, which explicitly have incorporated this philosophy in their stated reasons for existence. It is also true of BNOC, the British National Oil Corporation, and it certainly became true of ENI even before the death of Enrico Mattei. To many members of the political sectors in those countries the idea of profit is wrong, immoral. The mass nationalizations which took place in the United Kingdom were largely inspired by the distaste for profits which characterizes the philosophy of the Labour party.

There are state oil companies, however, that were created with a strong profit objective. These are the ones belonging to oil-producing and -exporting countries, essentially those belonging to OPEC. That these state oil companies should have a dominant profit objective is logical and understandable, since the economic, social, and political well-being of their shareholders depend, almost entirely, on oil. Whereas oil is an important but subordinate factor in British or Italian economy, for example, it means everything to Kuwait or Qatar and almost everything to Venezuela.

But even for these companies, the tendency to pursue social objectives has been very strong. This is more clearly observed in those countries which have a pluralistic political system.

Some of the social objectives which seem to be characteristic of, or have been imposed on, these companies, are providing direct economic aid to the communities where they operate; selling their products in the domestic market at a loss; utilizing locally manufactured goods and local services, even if they are not of the best quality or the most efficient; and serving as the catalyst of "national development," a grand but usually vague concept. Reflections on the latter goal moved Lord Beeching, the former chairman of British Railways to say:

> In the case of private industry there is a single, clear and unchanging primary objective. . . . In the case of a nationalized industry, on the other hand, objectives are more numerous, more ambiguous. . . . Moreover, they fluctuate in their supposed order of priority, not merely from government to government, not even from year to year, but almost from day to day at the whim of public and parliamentary opinion. . . . Management does not have the freedom to optimize its own performance in pursuit of a single objective, or even, in pursuit of a number of stable and compatible ones.[35]

Other characteristic social objectives are the development of backward or economically depressed areas or industries and the hiring of unemployed labor or technical staff regardless of their ability. Examples of each of these characteristics can be found in state oil companies from Third World countries or from industrialized countries. Petróleos de Venezuela's being ordered by the government to move the headquarters of two of its operating companies to the interior of the country, presumably to stimulate the local economy of those areas, is one example. ENI was ordered by the Italian government in 1971 to invest in southern Italy, but though "The government claimed that there was a need to create thousands of new jobs. . . . Subsequent events proved that excess capacity was created by this new investment and serious losses were incurred for both ENI and the government."[36]

During 1981 a decree was issued in Venezuela to enforce the buying of locally made products. For several years public officials in Venezuela have been ordered to travel only on the state-owned airline, an order which is not always followed. When Air France had to replace the obsolete Caravelles, the French government tried to force the company to buy a French jet, the Mercury 100, although this plane was "too big and expensive and . . . no other company in the world wanted (it)."[37]

In Mexico PEMEX has been obliged for years to sell its refined products to the domestic market at prices just a fraction of the international market price. The same has been true of Venezuela, where PDVSA has had to subsidize local industry and motorists in amounts exceeding $250 million per year. The subsidies, as economist have noted, frequently favor significantly only the big consumers, usually private multinational firms. In Perú Petroperú was long subjected to similar constraints, which sent the company's finances into chaos. The domestic price of gasoline in Perú in 1975 was fully 10 times lower than the refining cost. It was no wonder that by 1979 Petroperú had gone bankrupt and its debts had to be written off by the government. Essentially the same problem existed in Argentina, where state-owned enterprises were provided with fuel by the state oil company but frequently neglected to pay for it.

The case of Pertamina engaging in a multitude of nonoil ventures closely resembles that of YPF in Argentina and, in many instances, that of PEMEX and ENI. So far Petróleos de Venezuela has escaped dangerous diversification, although there have been strong pressures on this company to enter the agricultural, cattle-raising, and steel industries.

The Quest for Personal Power among Civil Servants

The work of David Coombes analyzed in detail the role played by ministers in the nationalized British industries. He says, "The distinction recommended by Herbert Morrison between general policy and day-to-day administration has never been maintained in practice."[38] He quotes a memorandum of the National Coal Board as follows:

> Through the attention they pay to investigate programs and projects, ministers . . . are able to involve themselves in almost as much of the detail of management as they wish, and to influence directly the operations of the (nationalized) industries.

The same memorandum on ministerial control (report from the Select Committee on Nationalized Industries) claims that "Ministerial interference with prices had made nonsense of this board's financial objective" (pp. 226–228). The money for the National British industries comes from the Exchequer, which explain the high degree of interference from ministers in the day-to-day operations of the industries. The anguish of the British over this political interference of their industries prompted the Select Committee to say:

> If a board is not meant to (have their company) behave as a private firm. . . If the men on the boards of the nationalized industries are not businessmen, how are the qualities of enterprise, initiative, and competent industrial management meant to be provided?

Coombes summarizes the role of the ministers in the nationalized British industries as entailing the following responsibilities:

> Appointing the members of the board and fix their salaries and pensions
>
> Giving general directions to the boards in matters of "national interest"
>
> Approving general programs and proposals involving large capital expenditures
>
> Authorizing borrowing and the utilization of surplus revenues;
>
> Authorizing diversification of activities, developing properties, pension funds

It is easy to conclude that bureaucrats having such a wide range of powers are in an excellent position to aspire to manage the industries they should only oversee.

The cases we have described in this chapter suggest that most state oil companies receive an inordinate share of attention from their contact ministers. CFP, ENI, Petramina, Petrobrás, and lately, Petróleos de Venezuela have been abundantly exposed to political contamination brought about by the activities of ministers. In the specific case of Venezuela, we have seen how the restraint of Minister Valentín Hernández (1975–1979) was replaced by the markedly political stance of Minister

Humberto Calderón Berti (1979–1983) and how the latter's systematic interference seemed to bring about a deep deterioration of the nationalized oil industry. In instances the behavior of public officers carries an important component of self-interest. Mexico's serious financial crisis of 1982 should at least partially be blamed on the selfishness and lack of courage of the top political actors shaping the country's oil policy. Accounts of the misfortunes of Petrobrás usually mention political infighting and personal power-seeking as two of the main ingredients of the problems. The fact that the president of Petrobrás in 1982 (an ideally technocratic position) had previously been Brazil's oil minister (an essentially political position) certainly deserves careful analysis. In Venezuela we have already seen how Oil Minister Calderon Berti was openly bidding for the job of president of Petróleos de Venezuela while still Energy and Mines minister, during the years 1981 and 1982.

Anthony Downs once suggested, somewhat pessimistically, that political officers tend to behave "solely to attain the income, prestige and power which come from being in office."[39] This assessment is not altogether unrealistic. In this sense few industries can provide politicians with a better path to personal power than the oil industry, especially in decades of almost continuous worldwide energy crisis.

Political Appointments and Cronyism

The appointment of friends, professional politicians, and generally incompetent personnel to highly paid positions has long been one of the common traits of state oil companies. The practice is encouraged by profit margins usually so great that managerial inefficiency tends to be masked much more easily than in other industrial activities. The proliferation of cronyism and political appointments, however, have often accomplished the antimiracle of condemning many state oil enterprices to chronic financial losses. In Argentina "the internal structure of YPF . . . showed the scars of the earlier politicization of the enterprise: some manifestly unsuitable people had been appointed to senior positions . . . personnel increased from 33,615 in 1970 . . . to some 53,000 at the time of the coup of April 1976."[40] In Perú appointments to Petroperú during the presidency of Velasco, were largely based on personal friendship. In Bolivia Yacimientos Petrolíferos Fiscales Bolivianos (YPFB) was used "to finance projects which are totally alien to the commercial nature of the agency . . . the employment of personnel for political reasons . . . In 1968, YPFB employed 4,200 staff (for a production of 8,000 barrels per day) as against 200 employed by Gulf Oil (for a production of 33,000 barrels per day).[41] G. Philip also quotes Bolivian oil workers complaining about "Ministers (who) change constantly, and each of them brings in his own people to 'work.' "[42] The case of PEMEX's job-selling, employment of friends and relatives, and the inordinate laxity concerning evident cases of conflict of interest has been abundantly documented in the works of Philip, G. Grayson, and E.J. Williams.[43] The problems of Petrobrás in this connection have been described in detail by P.S. Smith.[44]

Several instances of cronyism at high managerial levels can also be found in the case of ENI. Franco Grassini mentions the filling of positions for political patronage, commenting "Top management positions are filled by political appointment. . . . Since 1975, the political affiliations of management (in ENI) have become increasingly important."[45]

Petróleos de Venezuela was essentially free from this particular problem up to 1981 when the new board, structured in August of that year, included several members who clearly seemed to have been chosen for their personal friendship with Minister Calderón Berti. This move served to reinforce the influence of the minister in the day-to-day management of the Venezuelan oil industry.

The Desire for Government Control

Another common pattern to most, if not all, of state oil companies has been the desire of governments for incremental control of its activities. The perception held by the state bureaucracy and the political sector of national oil companies as "states within the state" has been very strong in Mexico, Brazil, the United Kingdom, Italy, Indonesia, and recently Venezuela. This perception has bred distrust and stimulated government efforts to minimize the power, real or imaginary, of these enterprises. In Venezuela the efforts have included frequent congressional hearings in which top oil executives are confronted with the doubts and the frequent criticism of the political sector. A change in the bylaws of Petróleos de Venezuela in 1981 obliged this company to submit its budgets to the Oil and Energy minister for previous analysis and approval. A more recent decision, made in September, 1982, put the oil-industry monetary fund in the hands of Venezuela's Central Bank. Although the apparent reason for such a move was the centralizing of all Venezuelan foreign reserves under a single organism, one of its immediate effects was that of diluting the decision-making power of Petróleos de Venezuela. In a similar fashion PEMEX's evolution has been marked by frequent attempts at control from the powerful workers' union, and from the federal government: "Mexican presidents exert power over the bureaucracy by selecting men and determining priorities rather than by detailed policymaking."[46] In Indonesia government control of Pertamina was increased by means of law 8, issued in 1971. The cash allowed to remain in the hands of this company was greatly restricted, in an effort to diminish Pertamina's financial and political power.[47] Law 8 also dictated placing Pertamina under the supervision of a board of directors made up of several cabinet members. This group could hire or dismiss managers of the company and had ample powers of control.

In general the desire for government control is a relict sentiment, derived from the fear which large multinational oil corporations have traditionally inspired in governments. Once these corporations no longer play a vital role in the domestic scene, the distrust and animosity of the government bureaucracy are seemingly transferred to the national oil companies, which, as in the case of Venezuela, have inherited much of the management philosophy, the procedures, and perhaps some of the secrecy and arrogance that characterized the style of private corporations. The desire for control is frequently fueled by the fear on the part of the political sector that the oil technocracy will become politically inclined and acquire more power than the politicians themselves.

Notes

1. L. E. Grayson, *National Oil Companies* (New York: John Wiley and Son, 1981), p. 12.

2. H. Madelin, *Oil and Politics* (Lexington, Mass.: D.C. Heath, Lexington Books, 1975).

3. Grayson, National Oil Companies, pp. 48–74.

4. Ibid.

5. Ibid., p. 52.

6. For a theoretical analysis of these factors, see J.P. Anastassopoulos, "The Strategic Autonomy of Government-Controlled Enterprises Operating in a Competitive Economy," Ph.D. diss., Columbia University, New York, 1973.

7. Grayson, *National Oil Companies,* p. 72. appendix 9.

8. Franco Grassini, "The Italian Enterprises, The Political Constraints," in *State-Owned Enterprise in the Western Economies,* edited by R. Vernon and Y. Aharoni (New York: St. Martin's Press, 1981), p. 74.

9. *Grayson, National Oil Companies,* p. 109.

10. Grassini, "The Italian Enterprises," p. 79.

11. Ibid.

12. Grayson, *National Oil Companies,* p. 123.

13. Ibid., p. 138, appendix 10.

14. G. Philip, *Oil and Politics in Latin America* (Cambridge Univ. press, 1982), p. 368.

15. Ibid., p. 374.

16. For a detailed account of the progress of politicization in

Petrobrás, see P.S. Smith, *Oil and Politics in Modern Brazil* (Toronto: McMillan, 1976).

17. Philips, *Oil and Politics in Latin America,* p. 377.

18. Ibid., p. 385.

19. Smith, *Oil and Politics in Modern Brazil.*

20. Philip, *Oil and Politics in Latin America,* p. 222.

21. Ibid., p. 225.

22. Ibid., p. 226.

23. G.W. Grayson, *The Politics of Mexican Oil* (Pittsburgh, Pa.: Pittsburg University Press, 1980).

24. Ibid., p. 20.

25. *New York Times,* January 6, 1940, p. 5, quoted by G.W. Grayson, ibid., p. 21.

26. G.W. Grayson, *Politics of Mexican Oil,* p. 333.

27. Ibid., p. 38.

28. Ibid., p. 51.

29. *Fortune,* May 4, 1981, p. 65.

30. G.W. Grayson, *Politics of Mexican Oil,* p. 120.

31. Philip, *Oil and Politics in Latin America,* p. 363.

32. Ibid., p. 367.

33. For a more detailed analysis of the organization of Pertamina, see R. Fabrikant, "Pertamina, a National Oil Company in a Developing Country," *Texas International Law Journal* 10 (1975):495–536.

34. Ibid.

35. In "Nationalized Industries, A Commentary," quoted by L. Narain, *Principles and Practices of a Public Enterprise Management"* (New Delhi: S. Chand, 1980).

36. Grassini, "The Italian Enterprises."

37. J.P. Anastassopoulos, "The French Experience: Conflicts with Government," in *State-Owned Enterprises in the Western Economies,* edited by R. Vernon and Y. Aharoni (New York: St. Martin's Press, 1981), p. 101.

38. D. Coombes, *State Enterprise, Business or Politics* (London: Alden and Mowbray, 1971), chapter 6, p. 87 ff.

39. A. Downs, *An Economic Theory of Democracy* (New York: Harper & Row, 1957), p. 28.

40. Philip, *Oil and Politics in Latin America,* pp. 420, 423.

41. Ibid., p. 460.

42. Ibid., p. 461.

43. Ibid.; G.W. Grayson, *Politics of Mexican Oil;* E.J. Williams, *The Rebirth of the Mexican Petroleum Industry* (Lexington, Mass.: D.C. Heath, Lexington Books, 1979), p. 101 ff.

44. Smith, *Oil and Politics in Modern Brazil.*

45. Grassini, "The Italian Enterprises," pp. 72, 74.

46. Philip, *Oil and Politics in Latin America,* p. 363.

47. Pertamina had truly become a state within the state. The Indonesian army, for example, was on the company payroll.

14 Management of State Oil Companies: The Venezuelan Case

The Venezuelan case is a good example of a nationalized oil industry possessing, from the beginning, a very strong managerial class. The reason for this was clear: Venezuela did not nationalize its oil industry in the early stages of evolution but in full maturity. Although a long period of time went by (1912–1933) before the first group of Venezuelans went abroad to be educated in oil-related matters and many more years would lapse until the government imposed the Venezuelanization of the oil-industry staff (1956), the country had ample time to form numerous groups of oil experts and managers by the time it finally took over its oil industry in 1975. By then the majority of oil-company board members were Venezuelans and a solid oil technocracy existed in the country. The depth of managerial and technical talent and expertise existing in the Venezuelan oil industry also helps to explain why the organizational model chosen for the nationalized enterprises was that of several operating oil companies working under the general guidance of a small financial- and planning-oriented holding company. If the resources had been leaner, the model would probably have been that of a state-owned, single oil company, a model which has not worked very successfully in other countries.

The political sector of Venezuela deserves much credit for recognizing the vital role that the Venezuelan oil technocracy had to play if the nationalization of the industry was to be successful. At no time during the analyses made of the possible options for nationalization was there any serious suggestion that the management of the industry should be in hands other than those of the oil-industry technocracy. Whether this recognition was based on conviction or in more pragmatic, circumstantial considerations is less clear in the light of more recent events. These events indicate the possible existence of a deep cleavage between those politicians who really believe in the technocratic role and those who feel that management of the oil industry can and should be transferred into political hands.

Even in a strong technocratic environment such as the one prevailing in the Venezuelan oil industry there has been a noticeable strengthening of the political factors in decision-making and strategy formulation. The political ingredient has been equally strong or stronger in many of the other Latin American state oil companies where professional managers or technocrats have been increasingly replaced by political appointees, by friends of the contact minister, or by protégés of the president himself.

Why has this happened? Is it the result of an inevitable process? Could evolution have been different? If so, how? These are some of the questions that should be addressed if a clearer understanding of managerial processes in state oil companies is to be obtained.

The Professional Manager

A brief, adequate definition of professional management was given by General R. Alfonzo Ravard, president of Petróleos de Venezuela, in a talk to the Institute for Advanced Studies of Administration of Caracas (IESA) in April 1981 at a time in which political interference in the management of the Venezuelan oil industry was already significant. Said General Alfonzo:

> In the 1950s . . . it became important (in Venezuela) to form specialized managers with the responsibility to conduct the big companies which were starting to be created . . . without having to be their owners. This is what has been referred to as professional management . . . the only (type) which can possibly be utilized in state-owned enterprises. . . .
>
> It is through the application of permanent and valid/managerial principles, which require a proper attitude and deep conviction, that the professional manager can replace what in more traditional and family-owned companies derives from the interests of ownership.[1]

This description of training of managers given by an experienced professional manager mentions several ingredients which deserve further emphasis: specialized training, application of permanent and valid managerial principles, deep conviction about their role, no ownership of the enterprises involved. Professional managers take a long time to be formed. They are specialists in the art of handling complex organizations and in keeping people highly motivated and working for a common objective. Formal training is extremely important, but equally vital is experience. In the oil industry the average time required to form a professional manager is 15–20 years, including 10 years of technical specialized activities, some 3 years of tailor-made managerial training, and some 5–7 years of middle-management assignments. To bring in a political appointee or a military officer, no matter how brilliant, into a top managerial position, is to risk deeply negative results for the organization, since there is no way to obtain knowledge rapidly enough of the procedures, and subordinates, and colleagues.

Professional managers do not make up their own rules. They work within a set of existing rules and procedures which change slowly in time as a result of the collective experience of the entire organization. There is no reason for professional managers to invent unique solutions, since in all probability the solutions already exist and have been successfully tested in the past. The nonprofessional manager, through ignorance of existing norms, procedures, and proven managerial tools, tends to improvise solutions, often becoming a source of disharmony within the organization, since organizations seem to respond best to coherent, predictable moves.

Professional managers also tend to have very strong opinions about their roles and believe deeply that what they do is in the interest of their organizations and in the public interest. Hence they are prone to oppose most forms of outside, unprofessional intervention. Many of these men and women will insist in participating in the decision-making process, especially since they will be called to implement such decisions. According to an interview with a Belgian executive made by R. Mazzolini, "A man who has made it to this level is usually a strong individual; he will not accept merely carrying out a decision without having participated in the decision-making itself."[2] The nonprofessional manager, having a lesser sense of tradition, often tends to be more accommodating, less rigid. Pragmatism and survival may be of higher priority to the nonprofessional than adherence to principle. At the same time professional managers "view themselves as expendable. . . . The readiness to put their jobs on the line over an issue they deem fundamental to their organization's long-range welfare is perhaps the simplest touchstone by which (they) are distinguishable. According to J. Bailey, "Men less sure of themselves—those . . . unable to face with equanimity the loss or surrender of their job with its power, prominence, and inward gratification—are generally tempted to rationalize away the importance of making such a . . . decision."[3]

Professional managers also see themselves as their own main judges for excellence and honesty. Bailey says, "a near majority of the top men in my sample (answered): I must satisfy myself, above all, that I have done the best I can do."

The fact that professional managers do not own part or the whole of the enterprise they work for allows them to concentrate on managing and lowers the possibilities that their actions might be unduly influenced by self-interest instead of the public interest or the interests of their institution. According to K.R. Andrews the manager's "share in profit is not his main motivation. . . . His satisfaction in increased compensation appears to be of considerable symbolic value in providing recognition of attainment and performance."[4]

In academic literature the professional manager has been characterized by the five criteria: knowledge (the subject of disciplined analysis); competent application (the existence of judgment and skill); social responsibility (not self-interest); self-control (to abide by a set of moral standards); and community sanction (the respect of society).[5] These criteria are actually very similar to those listed by Alfonzo Ravard from his more empirical vantage point.

Professional Managers and Professional Politicians

As early as 1918 M. Weber had already noticed "the emergency of modern bureaucracy—most especially the growing state apparatus, increasingly led by technically trained, professional career administrators."[6] Weber also had perceived "the rise of a new class of professional politicians. He was convinced "that inexorable historical tendencies would make this century of the professional party politician and of the professional state bureaucrat."[7]

Aberbach, Putnam, and Rockman define the clash of these two groups as "the axial problem of modern society."[8] The authors list a number of basic differences between the two groups. Professional administrators (bureaucrats) tend to have a more formal education. Politicians more frequently have a working-class origin, whereas bureaucrats tend to derive from upper-middle-class families. Bureaucrats tend "to come from families with a tradition of government service." They have usually reached their positions because they "have impressed their superiors, whereas the politicians got there by being popular and impressing their peers and constituents." In short, politicians are elected and professional administrators or managers are selected. Bureaucrats are "endurance runners" whereas the politicians are "sprinters." Bureaucrats seek technically appropriate solutions, while politicians define problems chiefly in terms of political principle and political advantage.[9] The likely attitudes of professional politicians at the helm of big corporations probably differ from those of the professional manager in the same degree that their backgrounds and paths to the top differ. Professional politicians tend to be guided by short-term political considerations since much of their power and even survival often depend on the impact they can produce on the electorate. That only a minority of politicians tend to behave with integrity is suggested by J.L. Fleishman when he says, "In most generations, however, we are fortunate enough to have a few politicians who are morally superior, who will under no circumstances tell a lie or decide an issue on the basis of how it will affect their career."[10]

The main weakness of the professional politician in a managerial position is often that of sacrificing the public interest to self-interest or to the interests of a small group or tribe. This is why actions by professional politicians should be evaluated on the basis of the interest they seem to serve. D.P. Warwick distinguishes four basic types of interests: the public interest, the interests of the constituency, the bureaucratic interest, and the personal interest.[11] It would seem clear that, for a public official to act in the interest of a group or, even worse, in his or her personal interest at the expense of the public interest constitutes an abuse of power and a violation of what Warwick calls the "ethics of discretion." Warwick lists five principles which should be followed by public officials in order to minimize the risk of ethical violations: (1) The exercise of discretion should, on balance, serve the public interest. (2) Reflective choices should be made, so that public officials can be clear about the values to be promoted or protected, rather than embrace them without examination. (3) Truthfulness should be exercised in the discharge of official responsibilities. (4) There should be respect for rules, standards, and procedures. "Official whim," warns Warwick, "is the enemy of a civilized order." Finally (5), there should always be exercise of restraint on the means chosen to accomplish organizational ends.[12] As an example of how not to behave, Warwick quotes John Dean's *Blind Ambition:*

> I soon learned that to make my way upward, into a position of confidence and influence, I had to travel downward through factional power plays, corruption and finally outright crime. Although I would be rewarded for diligence, true advancement would come from doing those things which built a common bond of trust—or guilt—between me and my superiors.[13]

Warwick' five principles can be utilized to evaluate the decision made by the oil minister of Venezuela to move the headquarters of two companies of PDVSA away from the capital in August 1981.

1. Was the decision for the public interest? It is difficult to say, since the decision was made without a proper analysis of the possible consequences.
2. Was it the product of reflective choice? It probably was not; the decision seemed to be made in a reactive mood rather than in a reflective fashion.
3. Was there truthfulness in the motivations given as the basis for the decision? Only time will tell, but the expressed main motive, that of a policy of decentralization of oil-industry activities, seems to have been forgotten.
4. Was there respect for the rules, standards, and procedures? The fact that management of the companies involved in the decision was not consulted and did not participate in it strongly suggests otherwise.
5. Was restraint exercised? To alter the lives of hundreds of employees and their families through a decision which was not duly analyzed indicates a clear lack of restraint on the means utilized to reach it.

It would seem therefore that this particular decision significantly departed from the set of ethical criteria proposed by Warwick.

Performance of the Main Players

The evolution of Petróleos de Venezuela clearly shows how its top managers were increasingly exposed to external pressures. When this happened, what was their conduct? Did significant changes take place in their attitudes? Let us briefly consider who the main players are.[14]

First, the top management players included the president of Petróleos de Venezuela, the presidents and vice-presidents of the operating companies, the members of the board of Petróleos de Venezuela, and the members of the boards of the operating and other subsidiary companies—together a group of some 60–70 persons. The executives immediately below top management included the coordinators of Petróleos de Venezuela, the senior managers of the operating companies, and what Mazzolini defines as key staffers, those persons very close to top executives but without line authority.

The top government players were the president of the country, the minister of Energy and Mines, other ministers of the cabinet, and high-ranking bureaucrats. Top political players included main leaders of the political parties, congressional leaders, heads of labor unions, and heads of business associations such as Fedecámaras and Pro-Venezuela.

Outside players were the professional societies, pressure groups of environmentalists, and organizations such as AGROPET, the association of oil-industry employees, and other ad-hoc players such as management consultants.

Players, according to R. Mazzolini, tend to behave in a predictable manner:

> a player's stand in one game takes into account his posture in the other games in which he is involved. . . . Rather, a player acts politically, that is, he does what other games he is involved in require him to do. . . . A minister responsible for regional development will push for investments by government-controlled enterprises in certain areas, but more often than not, not in terms of what is objectively most effective in view of the long-run interest of the collectivity . . . but in relation, say, to deadlines he himself faces—an election, for example.[15]

What is sought, suggests Mazzolini, are flashy, prompt results rather than long-term real achievement.

The Top Management Players

When confronted with increasing politicization and encroachment of the political bureaucracy, the president of Petróleos de Venezuela reacted by complaining internally within his own board of directors and by arguing with the political players against what he felt were increasing signs of deterioration. He presumably talked to the president of the republic, certainly talked to the contact minister and, more informally, exerted pressure on other players such as business leaders, political party heads, and so on, to enlist them against the wave of politicization. Finally he tried to postpone the execution of decisions made at the political level with which he was not in agreement, hoping that postponement would eventually mean cancellation. This seemed to be the case with the decision, already discussed, to move two subsidiaries of PDVSA away from Caracas.

The efforts of the president of Petróleos de Venezuela were not very successful. When a player does not prevail, there are several possible explanations. The player's opinion may not carry enough weight. Or although the player is highly respected, his or her opinions may be in the minority. Alternatively, the player's opinion may not be expressed with sufficient force and conviction to make an impression, or the player, although publicly against the decisions, may be privately in their favor. It is highly unlikely that the first, second, or fourth alternatives could have been the right ones. The opinion of Alfonzo Ravard had always carried tremendous weight in Venezuela. It was also widely known that a large majority of the players were against politicization of the oil industry. Alfonzo Ravard was also quite clearly against the attempts of the government at politicizing the industry. The most probable reason was therefore that he was not sufficiently committed. And why would this be so? Perhaps it is because he had other interests which had to be satisfied. He wanted to remain in power. Too strong a commitment would have led to a confrontation with the government and to his probable dismissal. When such a confrontation takes place, the government is almost always the strongest party. The PEMEX case offers at least two good examples of these power plays at the top. In 1976 Dovalí Jaimes accepted in silence the imposition of the labor unions and the government to unionize the technical staff of the company because he wanted to keep his job. Although the technocrats were solidly behind him, he chose not to make a stand. On the other hand Díaz Serrano paid with his job in 1979 for his deep difference of opinion with President López Portillo over oil-export prices, even though they were very close personal friends. This incident no doubt inspired Venezuelan Oil Minister Calderón Berti to remain silent when the Venezuelan government decided, in September 1982, to take the oil-investment fund away from PDVSA.[16]

It seems highly probable, therefore, that the president of Petróleos de Venezuela did not consider the issues serious enough to oblige him to make a sufficiently vigorous stand. The rest of the group of some seventy top managers behaved in line with their leader. They accepted the increasing signs of politicization with great restraint and discipline, a few of them showing personal anguish bordering on open rebellion. Maraven's president, A. Quirós, published a series of articles in *El Nacional, El Diario de Caracas,* and *Resumen,* which carried his vehement protest against the way the oil industry was being handled. As vice-president of Meneven I was dismissed over my protests against the move of my company's headquarters. Apart from these more overt signs of dissidence, there was little if any protest from this group. The explanation why this was so is not easy, since there might be multiple reasons for the attitude of several dozens of highly seasoned top executives. Many of these managers were highly honorable and dedicated men. A number of them were near retirement and wished to do so without conflict. A few of them were at least in partial accord with the new trends imposed by government. Some others had developed such strong tastes for personal power and economic privilege that they were not prepared to give them up for what they now considered to be rather fuzzy questions of principle. Still others had acquired a deeply ingrained sense of obedience after many years of service in highly organized, rather conservative enterprises. Many of the men who had made a life career in Exxon, Mobil, Texas, Gulf, or Shell were not passionate individuals but detached, practical businessmen who saw their duty in disciplined and efficient behavior rather than in controversy and protest. This attitude had served them and their organizations well for a great number of years. They saw no real need to change it, although most of them probably perceived that the atmosphere of the industry after nationalization contained ingredients which called for change. When to change and how much to change were questions still unresolved in their minds, however. What many of them tried to do was to express their views within the inner walls of the companies they worked for. Their strongest guide was the attitude of Alfonzo Ravard. As long as he did not rebel, they did not feel it was proper to rebel. They were highly conditioned to let their leader speak on their behalf. The feeling of obedience ran so strongly in this group that even those who wanted to use it as a mere rationalization for their lack of courage could do so without undue risk of being criticized by their peers or subordinates. For the many others who were very honest and felt the strong need to voice their concerns openly, it must have been a personal tragedy to see how their industry crumbled down, day by day, without their really being able to do much about it. Internal protest did not work, because all too often it became an incestuous exercise, a simple case of preaching to the converted, a momentary personal catharsis with no lasting, measurable effect in the diseased organizations.

Besides, they knew that the next step to ineffectual internal dissent was public dissent, what is popularly called "blowing the whistle." An essay by Sissela Bok gives us an academic insight into the more theoretical aspects of this often heroic, always controversial, action.[17]

As Bok mentions, the U.S. code of ethics for government servants asks them "to put loyalty to the highest moral principles and to country above loyalty to persons, party, or government department." Very few people could be found to disagree with such a code. Yet in Venezuela as well as in the United States, whistle blowers are exposed to reactions that range widely from accusations of disloyalty to charges of being mentally ill. The actions very often results in the immediate dismissal of the employee or in the best of cases in total loss of acceptance within the organization. It is not surprising, therefore, that most whistle blowers decide to break away at the moment of making their decision.

Reluctance to blow the whistle must have to do with the intuition on the part of the employee that his or her action will not be effective, that somehow the sacrifice will have been in vain. Bok reinforces this feeling when she says that "most whistle blowing, once undertaken, is destined to fail." There are several reasons why this is so.

The first reason is insufficient or no response from the audience. In Venezuela most people know very well that no scandal can hold public attention for more than a few days. Scandals erupt in the front pages of the newspapers and rapidly find their way into the back pages as saucier, more spectacular, fresher events take their place in the main ring of the political and social circus. It is very difficult, for a man who has built a career through many long years of hard and dedicated work, to be willing to risk it in a move that will at best put him on the spotlight for just a few hours.

The second reason is that, too often, solid documentation of the claims is hard or impossible to find. Even the closest friends of the whistle blower will refrain from providing information for fear that they might also fall with the dissenter.

A third reason is that the notion of public interest can be extremely hard to define. Because of this, many people tend to doubt the motivations of whistle blowers and to consider them as mere publicity-seekers or mudrakers. Although this tendency, according to Bok, will be minimized when the accusation is an open one and the accusing individual is clearly identified, the suspicion of personal gain or personal revenge is ever present.

Still another reason why denunciation is often a futile and resented action is that, once done, it pressures other people to take a stand. When the high executive of a company decides to come out saying that something is improper about his company, he inevitably puts pressure on his colleagues, either to support or to rebuke him. In most cases the uneasy compromise is silence, as even the most sympathetic colleagues will deeply resent being pressed into action. Furthermore, by corroborating the story of the whistle blower, they would almost be accepting their own tardiness in taking a similar action.

In the case of Petróleos de Venezuela, there is little doubt that no action was seriously contemplated by the group of top managers, especially in light of the absence of a decisive move by their leader, Alfonzo Ravard. Nor could any be reasonably expected.

Executives Immediately Below Top Management

The chances for protest from this group were much more remote. The coordinators of PDVSA, the managers of the operating companies, and the key staffers had to take their cues from their respective supervisors. Much more so than their leaders, they could not afford to take an individual stand. The coordinators of PDVSA were mostly specialists, hard-working technical staff highly concerned with their areas of responsibility. They did not feel at ease in the much more hostile environment of power plays, high political stakes, and public controversy—the arena of their superiors. The same can be said of the senior managers of the operating companies. Key staffers had to be even more cautious about their stand. These were people who owed their presence in the industry to a special relationship with the top managers. They were usually very competent people but also people who would not have been there unless they had a strong link with the man in power. In Petróleos de Venezuela these were people such as the personal advisor to the president, the legal advisor, and the executive secretary to the president. They knew probably better than anyone else what was going on, but their loyalty was clearly to the man they had been called to assist. If they spoke their minds, and there are plenty of reasons to assume they did, it was always on a one-to-one basis with their bosses. They did not build scenarios of insurgence. Their very reason for existence was to reinforce the position of the man in whom they essentially trusted. If there was any difference of opinion, they mentioned it to him, even passionately, but probably felt that mentioning it was the outer limit of their responsibility.

The Government Players

These players wanted to have a stricter control of decision-making within the oil industry. Up to 1979 the relationship between the president of the country and his oil minister had been one of mutual respect but without doubt in anyone's mind about who was the boss. The president was clearly in charge. After 1979 the relation between President Luis Herrera Campíns and his oil minister, Calderón Berti, was one of a close personal friendship and a substantial delegation of authority from Herrera to Calderón Berti. Although the conceptual elements of this relationship were all positive, in practice it meant that the minister could develop practically unchecked power over the industry. His early views on the necessity for an expanded role for himself and his ministry staff in policy-making slowly evolved into a desire for personal control, personal power. The changes in the bylaws of PDVSA, the naming of friends to key positions in OPEC and the board of PDVSA, decisions about management salaries and responsibilities, the decision about moving Meneven and Pequiven away from Caracas were all examples of his desire to increase personal control of Venezuela's main industry. What might have started as a genuine desire to establish a better balance between oil industry's autonomy and the government's right to be informed and to formulate basic policy seemed to become a quest for personal aggrandizement. This was possible because as Minister Calderón Berti tentatively probed top oil-industry management, he noticed that resistance to encroachment was highly diluted by objectives of survival and personal security. Finding little resistance from top management players, the minister started to develop a more ambitious personal scenario: to become president of PDVSA in 1983 and to use this position as a stepping stone to even more important positions in the 1990s.

In so doing, however, he paid a high price. He lost credibility within the oil-industry technocracy and elicited resistance from technocrats and politicians alike, the first group feeling deceived by the change in the minister's attitude, the second group feeling threatened. He also found opposition from other members of Herrera's cabinet, especially from the minister for the Venezuelan Investment Fund, L. Díaz Bruzuál, who later became head of the Venezuelan Central Bank.

In October 1982 Calderón Berti had to abandon all remaining pretenses of being on the side of the oil industry and against government intervention. Although the takeover by the Central Bank of the oil-industry investment fund was rejected by the oil-industry technocrats, by the main opposition party (Acción Democrática), and by much of public opinion which had a deep distrust of the motivations of the government, Calderón Berti chose to defend the government's action. The main spokesman on oil matters for Acción Democrática, Arturo Hernández Grisanti, publicly criticized him for not resigning after he had spontaneously promised to do so if the government did what it finally did.[18]

In general, therefore, the attitude of most top government players seems to have been characterized by the focus on rather narrow bureaucratic interests and on personal, power-seeking interests. Even the most selfless-looking decision, that of transferring the oil funds to the Central Bank, had a dominant short-term component since it was at least partially designed to allow the Herrera administration to end its term without having to devalue the national currency or to install exchange controls.

The high-ranking staff at the Ministry of Energy and Mines was primarily interested in expanding their power quota, which had been severely curtailed with the creation of PDVSA in 1975. We have seen how the efforts of this group during the first 5 years of nationalization led to major confrontations with the oil-industry technocrats and in general to inefficient handling of the day-to-day interaction between the industry and the government sector.

Top Political Players

The leading political figures of the country have always been very conscious of the importance of keeping the oil industry generating income at optimum efficiency. They know that without an efficient oil industry no government could obtain the tools for national development. At the same time there seems to exist in most of them an urge to curb what they perceive to be a threat to their own political power, that is, the growth of the national oil industry. This fear is quite widespread among the political sectors of countries having a large national oil company. A Canadian paper on Crown corporations clearly voices this concern:

> There is a certain amount of conventional wisdom surrounding Crown corporations performing activities on a commercial basis . . . to the effect that the government and Parliament must avoid all but the most cursory intervention into their affairs lest their commercial performance be jeopardized. This view ignores the fact that without exception such corporations were established by the Government of Canada to achieve broad policy objectives.[19]

The fear commonly expressed by politicians is that state oil companies grow too big and powerful to start to work for their own corporate interests at the expense of broader economic, social and political objectives. Although there is no doubt that this was the case of Pertamina under Dr. Sutowo, it is far from being the rule.

Some politicians view increasing control of state oil companies as a source of political power for themselves or their political organizations. In Venezuela the fight for control mostly responds to this sentiment. Consequently it is very strong and, although it is adversely influencing the efficiency of the industry and its income-generating capability, many political leaders seem to be prepared to pay this high price to harness oil-industry management into more docile obedience to political directives.

This desire is often fueled by a widespread antipathy among politicians toward oil technocrats. Petrobrás is a good case in point. For many years, the technical staff and the managers of the Brazilian state oil company have been seen as arrogant and secretive. The same accusation has been repeated in Indonesia, Mexico, Italy, and the United Kingdom. In Venezuela this antipathy is well marked and is due to several reasons: the relative high success of the oil industry combined with the consistent failure of other state enterprises generally run by political appointees; the high salaries and standard of living enjoyed by oil-industry executives; the perception that a link still exists between oil-industry executives and their former multinational colleagues; and the arrogance and air of superiority which the political sector perceives as traits of many oil-industry executives. These reasons mix ideology with more visceral feelings of distrust and even envy to form in an attitude of politicians toward oil technocrats which is at best neutral and at its worst strongly hostile. The president of one of the operating companies commented as follows: "No politician likes us. They want to see us walking along with the crowd, our heads down. When the Central Bank took the oil-investment fund away from PDVSA's direct supervision, I met a prominent COPEI political leader at a restaurant, and he told me, with glee: 'Now you are just like all the rest, no money.' "[20]

This wish to cut the industry down to the size of the often imperfect bureaucratic system of the rest of the state is a strong component of the attitude of top political players vis-à-vis the Venezuelan oil sector. In the National Congress this drive is expressed through frequent hearings in which oil-industry executives are occasionally harshly treated.[21]

Although the antipathy is very widespread, some political leaders do feel that there is a definite role to play for oil technocrats in the management of the industry and interact with them with moderation. Acción Democrática political leaders have shown themselves much more restrained in this respect, although this has not always been the case. Predictably, the most virulent attacks usually come from the representatives of small extremist leftist political parties, people who have made of these attacks almost a full-time job.

The heads of labor unions and of business groups play a much more supportive role in the oil industry. Labor unions have adopted a very moderate attitude in their relationship with the oil industry and do not share in the attacks of the political sectors. Their objectives do not include the control of the industry but better working conditions for their members. They have obtained that. Business leaders organized in Fedecámaras see the oil industry as the main stabilizing agency of the Venezuelan economy and generally tend to support their management, especially when led by a man such as Alfonzo Ravard, who has the full acceptance of the Venezuelan business community. An exception is R. Cervini, who has been president of Pro-Venezuela for several years. An ardent nationalist who likes to surround himself with radical intellectuals, Cervini advocates a strict government control of the oil industry and has publicly shown distrust of top oil-industry management.

Outside Players

In addition to all the individuals and groups already mentioned, there are several other players influencing management processes in Petróleos de Venezuela. One of the most important is the Venezuelan Engineering Society (Colegio de Ingenieros de Venezuela), the largest and most powerful professional society in Venezuela. This society is an advisory body to the government in professional engineering matters. Since the oil industry employs hundreds of engineers and engages in numerous engineering projects and megaprojects, it has received the special attention of the society. In the 1970s the board of the society became increasingly politicized, to the extent that election to the board is done in the name of the political parties (the president belongs to COPEI, the secretary to Acción Democrática, and so on). Because of this, the decisions of this board have frequently incorporated a substantial political component, and the positions of the organization on important oil-industry issues tend to reflect the positions of the political parties represented in the board. In general the society has advocated a strongly nationalistic stand in oil matters. In 1956 it pioneered the Venezuelanization of technical and managerial ranks of the oil industry. In 1971 it supported the law of reversion and in 1974 the nationalization of the oil industry. After nationalization the society pressed hard for increased participation of Venezuelan engineering firms in the planning, design, and execution of oil-industry projects and for the issuing of government regulations restricting the utilization of foreign engineering firms. They also objected to the continued practice by the oil industry of hiring foreign engineers, although they recognized that there was a chronic shortage of experienced engineers in the Venezuelan work market and allowed a restricted level of hiring to take place. The society was also influential in the issuing by government of government decree 1234, by means of which the oil industry had to utilize increasing amounts of Venezuelan goods and services (even if they were up to 20 percent more expensive than equivalent foreign ones). The decree stimulated the creation of national or mixed engineering companies to serve the oil industry, practically barring the utilization of fully owned subsidiaries of foreign engineering companies. In this endeavor, the society was enthusiastically supported by the Petroleum Chamber, a group of Venezuelan service and engineering companies which forcefully lobbied for increased participation in oil-industry activities. The Petroleum Chamber rapidly became a powerful pressure group with strong links to the Engineering Society and to important members of the state bureaucracy.

AGROPET, the group created in 1974 to serve as a vehicle for the opinions of concerned oil-industry technocrats continued in existence after nationalization, although in a more subdued role. Membership decreased from the more critical months of 1975, but still included a very important cross-section of the industry. As the group never became a white-collar union and preferred to emphasize the analysis and discussion of oil-industry issues, it has never become very popular, although its opinions usually had the ear of technocratic and political sectors alike. During the period 1979–1982 AGROPET was the only organization which systematically called the increasing politicization of the oil industry to the attention of public opinion. That their warnings were not very successful does not detract from their efforts but probably adds to the responsibility of the players who could have changed the course of events yet chose to remain silent and passive. In October and November 1982 AGROPET vigorously opposed the decision to transfer the oil-investment fund to the Venezuelan Central Bank and harshly criticized COPEI's congressman Haydeé López Acosta, who had publicly said that the minds of the oil-industry executives "still had to be nationalized," a favorite argument of flag-waving Venezuelan politicians.[22]

Is Involution Inevitable?

In most state oil companies the process of deterioration has started when the original objectives are abandoned in favor of less clear ones. The pattern seems to be marked by several stages:

1. The new company defines its objectives and managerial philosophy and starts working within these objectives and philosophy. This was the case with Petróleos de Venezuela in the period 1975–1979.
2. The original objectives, philosophy, and goals are progressively abandoned in favor of shorter-term political or social objectives. This stage appears the more rapidly in open, participative democracies as ministers and pressure groups, as R. Vernon suggests "can be expected to use its power to influence the firm's behavior."[23]
3. The interaction between the political sector and the top managerial hierarchy of the state oil company results in the progressive encroachment of political motives over profit-oriented, technocratic, managerial decision-making. This became very evident in Petróleos de Venezuela during the period 1979–1982.
4. Politicization travels downward throughout the organization as mismanagement, political appointments, and personal power-seeking become institutionalized and an accepted practice. The usual symptoms include job-selling, the erection of personal baronies, widespread conflict of interest, and small-scale to moderately widespread corruption.
5. Politicization is now complete. Motivation of personnel is very low. There is widespread corruption, absenteeism, and featherbedding. The

use of company resources to pay back political favors or to increase personal power is common practice. At this stage the company might be described as terminally ill.

These stages can be visualized as a continuous, overlapping process. Even if an organization could be described to be essentially at, say, stage 3, some elements of stages 4 or 5 could already be present in an incipient manner. In those cases the external observers tend to be largely insensitive to criticism of the firm by other outsiders or even by insiders unless the evidence is very clear. The firm will be described as in relatively good shape, especially as compared to other state-owned enterprises, which are in a more advanced stage of involution. Petróleos de Venezuela seems to be a stage 3 company showing some elements of stage 2 on the one side, and of stage 4 on the other.

These stages are not necessarily irreversible. One case which illustrates how an extremely deteriorated organization can be cleaned up and put back on its feet is the Venezuelan Petrochemical Company, Pequiven, now under the supervision of Petróleos de Venezuela. This was clearly a stage 5 company when it was ascribed to PDVSA in 1978. The strategy followed in its rehabilitation consisted of injecting at top managerial level a group of oil-industry managers who took with them the procedures and norms prevailing in the operating oil companies and who applied these procedures. In addition there was a very significant reduction of the labor force and a streamlining of operations. The main objective was set at getting the existing plants into normal production before even attempting renewed expansion. As a result Pequiven is now a stage 3 company together with the rest of the oil industry, except that the direction from which it came is radically different from that of the other affiliates of Petróleos de Venezuela. The example of Pequiven strongly suggests that the involution of state oil companies is not an inevitable process and that a reversal of the tendency toward deterioration is very much possible.

Notes

1. R. Alfonzo Ravard, "Petróleos de Venezuela, "Cinco Años de Normalidad Operativa," Caracas, 1981.

2. R. Mazzolini, *Government Controlled Enterprises: International Strategic and Policy Decisions* (New York: John Wiley and Sons, 1929), p. 286.

3. J. Bailey, "Clues for Success in the President's Job," in *Developing Executive Leaders,* edited by E. Bursk and T. Blodgett (Cambridge, Mass.: Harvard University Press, 1971), p. 67 ff.

4. K.R. Andrews, "Towards Professionalism in Business Management," chapter 1 in Bursk and Blodgett, eds., *Developing Executive Leaders,* p. 8.

5. Ibid., pp. 4–5.

6. In J. Aberbach, R.D. Putnam, and B.A. Rockman, *Bureaucrats and Politicians in Western Democracies* (Cambridge, Mass.: Harvard University Press, 1981), p. 1.

7. Ibid., p. 1.

8. Ibid., p. 238.

9. Ibid., pp. 240–244.

10. J.L. Fleishman, "Self-Interest and Political Integrity," in J.L. Fleishman, L. Liebman, and M. Moore, eds., *Public Duties: The Moral Obligation of Government Officials* (Cambridge, Mass.: Harvard University Press, 1982), p. 81.

11. D.P. Warwick, "The Ethics of Administrative Discretion," in Fleishman, Liebman, and Moore, eds., *Public Duties,* p. 112.

12. Ibid., pp. 115–123.

13. Ibid., p. 123.

14. For an excellent theoretical discussion on players and their roles, see Mazzolini, *Government Controlled Enterprises* chapter 9, pp. 283–310. I have adopted a modified version of Mazzolini' grouping of players.

15. Ibid., p. 309.

16. Although it is said that Calderón Berti was privately against this measure and that he had promised to resign if it was ever taken, the truth is that when the time came he publicly defended it. See *Resumen* 472, November 1982, pp. 13–20.

17. S. Bok, "Blowing the Whistle," in Fleishman, Liebman, and Moore, eds. *Public Duties,* chapter 8, pp. 205 ff.

18. *El Universal,* October 4, 1982, pp. 1–14.

19. Canada, Privy Council Office, "Crown Corporations: Direction, Control and Accountability," Ottawa, 1977, p. 21.

20. Private conversation, November 1982.

21. Congressman H. Pérez Marcano in a meeting in Congress on November 10, 1982 said, "I don't trust Petróleos de Venezuela. . . . I think Rodriguez Eraso or Quirós (presidents of Lagoven and Maraven) are puppets of the multinationals. I don't trust Alfonzo Ravard." in *El Diario de Caracas,* November 11, 1982, p. 3.

22. See *El Universal,* November 5, 1982, P. 1–12 and November 6, 1982, p. 2–2.

23. Vernon and Aharoni, eds. *State-Owned Enterprise in the Western Economies,* p. 12.

15 The Road Back to Efficiency

Involution does not have to be an irreversible process. In democratic countries a change in government often brings about a change in existing political, economic, and social attitudes, including those related to the philosophy of managing state-owned enterprises. Sometimes the change is for the worse, but it can also be for the better, as in the example of Pequiven's changing for the better after being taken over by Petróleos de Venezuela in 1978.

Petróleos de Venezuela inherited in 1975 a group of oil companies lacking those characteristics one tends to associate with state-owned enterprises. These companies were going concerns and had strong corporate identities and a clearly defined profit orientation. Some were better managed or more autonomous than others, but all, with the possible exception of Corporación Venezolana del Petróleo, CVP (the original state oil corporation), had a set of strong, well-established norms and procedures. It would take the nationalized companies about 5 years to experience a measurable degree of change, to acquire characteristics more in line with state-owned enterprises than with the multinationals from which they had derived. The starting point for this process of change in most enterprises is usually the replacement of the profit orientation with less clear social motives. When this initial change takes place and many of the members of the organizations are not in agreement, morale and motivation decrease and deterioration sets in. In the case of Petróleos de Venezuela, it is clear that a majority of the managerial and technical staff still feel strongly that profit, maximum rentability, should be the basic objective of the Venezuelan oil industry. At the political level and at the interface between the political and technocratic sectors, however, strong forces work to dilute this objective into others of a social nature. Some of these new objectives have already found their way into the planning documents of the industry, even though some of them would seem to be in collision with proven, long-established objectives such as maximum efficiency, minimum costs, and recruiting of the best candidates in the work market.[1]

As a result of these pressures Petróleos de Venezuela underwent a process of involution which seriously affected its capacity to operate with efficiency. To stop or reverse this process of involution, three basic steps

275

should be taken, all entirely feasible and realistic:

1. Establish a political pact specifically dealing with the oil industry.
2. Formulate basic guiding principles to be strictly adhered to.
3. Develop a clear modus operandi between the political and the technocratic sectors.

A Political Pact

One of the most useful tools of Venezuelan democracy has been the capacity and willingness shown by the main political parties to reach agreements in fundamental issues and to celebrate pacts. This was the case in 1958 after the collapse of the Pérez Jiménez dictatorship. The main political Venezuelan leaders—Betancourt, Caldera, and Villalba—met to discuss and to agree on political floors and ceilings.[2] As a result of these discussions, the political tensions which had reached excessive levels during the 1940s and which had lubricated Pérez Jiménez's access to power, were minimized, at least during the all-important initial period of democratic consolidation. The agreement was probably difficult to reach and must have required substantial restraint on the part of the men who worked it out, but they knew how vital it was in order to eliminate a repetition of the sad political experience that had just ended.

In the 1980s, a quarter-century later, the country again seems to be in great need of the capacity for agreement shown by the political leaders of 1958. Specifically, this is the case with the oil industry. There should be a political pact aimed at depoliticizing the oil industry.

Although it might sound odd and unrealistic to propose a political pact designed to exclude politics from the management of the most important Venezuelan industry this is in fact the only move which would benefit the industry, the country, and the political sector. The oil industry is the main, almost the only important source of revenue for the Venezuelan government. Without an efficient industry, this source of revenue would shrink and with it the hopes for a continued vigorous and healthy democratic system. A political pact as suggested would therefore be much less idealistic and much more pragmatic than might appear at first glance.

Such a pact would have to be public and explicit. Much of its value would lie on its being didactic. Public opinion would be shown the difference between political and economic, income-generating institutions and the virtues of keeping the latter outside the realm of the former.

A political pact with the industry should clearly state the commitment of the political sector to keep the oil industry free of political pressures. Comment, analysis, and criticism of the industry should be aimed at issues rather than at persons and should be institutionalized and expressed through the proper channels rather than aired indiscriminately. The political parties would instruct their members not to utilize the industry for the obtaining of partisan or personal benefit. A pact should emphasize the income-generating role to be played by the oil industry in preference to an income-distribution role, which is more the realm of government.

A political pact should recognize the role to be played by professional managers, paid by the state to conduct the affairs of the industry in accordance with policy guidelines formulated by policy-makers at the political level but with the input and participation of the managerial sector. This relationship should guarantee that the industry responds at all times to national priorities as defined by the legitimate representatives of the people.

A pact should make clear the necessity of full and precise information being provided by the industry to the representatives of the owner. One traditional area of friction between industry and government has been the issue of government control, expressed as desire from the side of the owners to keep the industry responding to their orientation and directives. This desire has often led to excesses fueled by distrust. Better information should establish enough trust to keep government free from the temptation to interfere. To control is to be properly informed so as to be sure that basic policy is being followed. At the same time the use of oil industry information by the political sector should recognize the confidential nature of much of the data and used with restraint. The misuse of information could rapidly lead to its interrupted flow and to the reappearance of jurisdictional conflicts.

A political pact with industry should recognize the necessity to compensate the managerial and technical staff of the industry competitively with the international market. It is difficult for the political sector to accept that there should be a difference in treatment between state bureaucrats and oil-industry employees, but this is one of the facts that have to be recognized as essential. The country cannot increase the rewards of all state employees to the level of the oil-industry employees, but it should not decrease the rewards of these employees to the same level as public administration employees. Competitive rewards for the employees of a vital, income-generating industry have to be viewed as an investment.

Finally, a political pact should include the commitment of all political sectors to analyze carefully and consult with each other before making major policy decisions in the field of hydrocarbons.

Basic Guiding Principles

Clearly defined basic guiding principles are essential to ensure the continued success of any enterprise. A mission has to be established, then objectives. Strategies can change in response to the changing environment, but the mission and the main objectives should have a much more permanent nature. In the case of the Venezuelan nationalized oil industry, the extensive national debate which took place for more than a year before the decision was finally made, gave all sectors enough time to distill their views and to agree on a group of principles which became clearly understood by all the main political and technocratic actors. These principles were self-financing, freedom from political interference, promotion of personnel based on the merit system, professional management of the industry, and normalcy of operations.

Self-Financing

That the national oil industry should be able to generate its own capital requirements was one of the earliest concepts to be agreed upon. It was the main concern of the president of Petróleos de Venezuela, General Alfonzo Ravard, to find a mechanism which would ensure that. The nationalization law incorporated an article, number 6, which reserved for Petróleos de Venezuela 10 percent of the net product of oil exports. In 1975 the amount of money going to Petróleos de Venezuela through this mechanism seemed sufficient. It soon became clear, however, that the rapid expansion of the industry would require additional sources of capital. The government agreed to leave within the industry the net profits of the operating companies and to establish an oil-investment fund, managed by the industry, to be utilized exclusively for oil-industry projects and development. Through these mechanisms Venezuela was making sure that the oil industry would not run into the difficulties so commonly experienced by state-owned oil companies in other countries. The case of PEMEX is illustrative. The Mexican oil company has been chronically incapable of financing its projects and has had to resort to short-term outside financing. By 1982 the external debt of PEMEX stood at around $25 billion and the enterprise could not do anything but to keep borrowing. It should be evident that no efficiency is possible if capital is not available when and how required. To be forced to go either through the international banking system or the political mill to obtain funds carries a very high risk of delays or cancellation of vital projects.

The Venezuelan oil industry was free from this danger. By early 1982 the oil-investment fund had reached about 8 billion. In September 1982, however, the economic conditions in the country had deteriorated so badly that the government decided to centralize all international reserves in the Venezuelan Central Bank. This move could not be disputed on technical or legal grounds but certainly had doubtful merits in the long term. It meant, among other things, that the managerial and financial autonomy of the oil industry was now on much less certain grounds. A further event, in December 1982, seemed to corroborate this fear when the government ordered Petróleos de Venezuela to utilize about $2 billion of the oil-investment funds to buy government bonds of the public debt. There is no doubt that such a move greatly contributed to the further loss of motivation among the oil-industry technocracy.

Freedom From Political Interference

The necessity for a political pact to eliminate the politicization of the national oil industry has already been mentioned. Politicization, according to the president of Petróleos de Venezuela, General Alfonzo Ravard is the modification of systems and procedures due to political circumstances, the contamination of these systems and procedures by temporary, external interests.[3] It is also, he said, the making of fundamental decisions about the oil industry based on elements of judgment which do not apply specifically to the industry. Unfortunately General Alfonzo Ravard has been much more successful in defining what politicization is than in resisting it.

Politicization is a somewhat generic term applied to different types of violation of healthy management practices. It has been applied to decision-making for political motives, to political appointments, to the pursuit of personal objectives by government actors, to nepotism, corruption, and a host of other ailments. What is perhaps really important is that its connotation is almost invariably bad. Not everybody sees it that way, however. The desire for control over the oil industry by certain sectors of Venezuelan political life has outweighed concern over politicization, to the extent that they seem to be prepared to pay this price for control. The conviction that a nonpoliticized oil industry is better for the country and for the politicians themselves than an industry subject to politically intervention is essential if this guiding principle is to be maintained and respected by all.

The Merit System

The principle that promotions and advancement within the organizations should be based on merit is not new and is now an accepted guideline of healthy organizations everwhere. Conversely the violations to this principle are an accepted practice of unsuccessful organizations. The extreme cases of violation of this principle are cronyism and nepotism, the naming of friends and relatives to positions within the organization regardless of their qualifications and of the merits of others. For many years the oil companies which were nationalized in 1975 utilized the merit system for their promotions. However, the naming of the board of Petróleos de Venezuela in 1981 introduced a clear change in this system, since three of the new members of the board were clearly placed there because of their friendship and personal loyalty to the minister whose recommendations were all-important to the decision.[4] The minister also announced, in 1982, that he would be a candidate for the presidency of Petróleos de Venezuela in August 1983. This announcement ran in clear violation of the systems traditionally applied by the industry, systems that he should have been the first to uphold. Adherence to the merit system is one principle relatively easy to comply with; it constitutes one of the most effective tools to maintain morale and motivation within the organizations.

Professional Management

The role of the professional manager in contrast to that of political appointments has been discussed in the preceding chapters. The cases of PEMEX, PETROBRÁS, PERTAMINA, and ENI show abundant examples of political appointees reaching key managerial levels in their organizations and the negative influence of those appointments on overall efficiency. Petróleos de Venezuela has been remarkably free from political appointments largely because of the depth of managerial talent within its organizations and the fact that the displacement of career personnel by political appointees is a form of attack which touches directly on people and tends therefore to elicit more vigorous resistance. As opposed to other state oil companies, the tradition of professional management in the companies of Petróleos de Venezuela has been both old and strong. PDVSA's case seems to suggest that the nationalization of already mature organizations has a better chance to be at least temporarily successful than the nationalization of an industry for which the organization has to be essentially created by the state after the decision to nationalize has taken place.

Normal Operations

An industry which provides the country with over 90 percent of its foreign exchange and is the sole motor of the national economy should not be the object of abrupt or even moderately planned-for operational interruptions. When countries, either developing, like Nigeria and Venezuela, or developed, like Norway and the United Kingdom, grow accustomed to receiving a daily oil paycheck, any interruption to this flow of easy money tends to produce a national economic crisis.[5]

The record of Petróleos de Venezuela in this respect has been excellent.

Even during the period 1976–1979 when the industry undertook a major organizational revamping which brought the number of operating companies down from fourteen to four, operations went on in an uninterrupted fashion. This was possible because the move was carefully planned and because there was much commitment and discipline on the part of oil-industry staff. In 1981, however, the political decision to take the headquarters of Pequiven and Meneven away from Caracas and to Maracaibo and Puerto La Cruz was received with much less commitment and even less discipline. As a result little progress was made during 1982, and, if anything, the move contributed to the deterioration of the internal cohesion of these two companies, reducing their general efficiency. A more delicate decision was made in 1982, when operating areas and personnel changed companies in an effort to continue the reorganization process which had started in 1976. Although the conceptual basis for this decision was valid, the timing was certainly very poor. In 1982 at least two companies, Corpoven and Meneven, were still experiencing important organizational and managerial adjustments as a result of the original reorganization. The superimposition of fresh changes onto the still-unsettled, older ones could not contribute to the improvement of those organizations. To the contrary, it added to the confusion and general lack of motivation of the staff.

The direct results of transferring whole segments of one company to another might well be loss of efficiency and greater costs per barrel. Personnel who have worked for one organization for a long time will now have to learn new ways and to make new friends.

Normalcy of operations is a principle not to be confused with immutable working conditions. Rather, it should convey the need for the careful planning of changes, for smooth transitions, and for excellent timing. It should also convey the need for participatory planning between the government sector and the industry, another concept systematically advocated by oil-industry management as a basic requirement for success.[6]

Coexistence of the Political and Technocratic Sectors

The development of a clear modus operandi between these sectors is the third and probably the more difficult step in the process of maintaining an efficient nationalized oil industry. The differences between the two groups are so vast and deep that much goodwill and understanding will be required to bring them together. Nobody would pretend that they should solve all their differences, but a more modest goal is that they defined their respective roles in policy-formulation and in the management of the oil industry. The starting point for this role-definition should be the mutual acceptance that such definition is both possible and desirable. Up to now this acceptance simply does not exist. There is too much distrust and antipathy between the two groups. This visceral stage will have to give way to a more rational and businesslike relationship. Again, love is not what is asked for, but civilized coexistence. The distrust and antipathy have deep roots which have been examined elsewhere in this book: the traditionally friendly relationship between Venezuelan dictators and oil-industry top management at times in which political leaders were in prison or in hiding; the training of the Venezuelan managers at the hands of the multinational oil corporations with all its ugly implications of lack of patriotism;[7] the perceived arrogance of oil-industry technocrats; the volubility and banality of political attitudes; the ignorance of the political sector about oil matters; the perceived shaky principles of politicians. The deterioration of the oil industry has reached the point, however, where both groups should clearly feel the need for coming together.

Should they try to do this, they might find that the essential issue will be that of the political sector accepting the existence of an oil-industry elite, not to weaken democracy, but to keep it strong. A main argument of the political sector regarding elites has traditionally been one of incompatibility with the democratic system. In Venezuela, political leaders have traditionally been wary of business and military elites on the grounds that they present a "threat to the democratic system." What seems to be the case is that politicians perceive those groups as a threat to the well-being of their own political elites rather than to the system. Since the late 1960s the tendency among the Venezuelan political elites, defined as the minority groups which have effective control of the major political parties, has been that of weakening other elites within the country so as to gain their control. This has been the case with the military, the professional societies, and with some of the business elites, and it explains the efforts now being done to curb the power of oil-industry management, which is also perceived as a potential political threat.

In Mexico, the political sector has never shown a noticeable distrust of the PEMEX elite because, as G.W. Grayson points out, Mexican oil-industry technocrats have never shown a real desire for political power.[8] But in Venezuela this distrust has been a real obstacle to peaceful coexistence.

The reason for this distrust also has to be found in the size of the country and the very limited number of members of its top elite. K. Deutsch estimates that the top elite in any Western-style democracy is about 50 members for 1 million people. In a country of 16 million people such as Venezuela, the top elite would only consist of about 800 people. Since about 100 of these members belong to the oil-industry elite, it is easy to see why the political elites should consider them as a potential threat. The members of the Venezuelan political elite would probably not exceed 300. The hard-core business elite numbers about 200 members and is largely sympathetic to the oil-industry technocracy. In a sense, therefore, the political elites feel surrounded by ideological enemies, a perception which tends to reinforce their defensive attitude.

It is clear, however, that unless the two groups come to accept each other and come to terms regarding each other's role and jurisdiction, there can be no hope of the national oil industry being efficient and, therefore, there will be no hope for a healthy Venezuelan democratic system.

Notes

1. "Buy Venezuelan," the hiring of unemployed engineers, and direct economic aid to communities where companies operate are some of these new objectives.

2. For a theoretical discussion of this concept, see R.A. Dahl, *Dilemmas of Pluralist Democracy* (New Haven, Conn.: Yale University Press, 1982), p. 170. For a detailed discussion of the Venezuelan experience, see J.A. Gil, *The Challenge of Venezuelan Democracy* (New Brunswick, N.J., 1981).

3. Alfonzo Ravard, *Cinco Años de Normalidad Operativa,* p. 234.

4. For a more detailed account of the 1981 events, see the *Wall Street Journal,* February 16, 1982: "Conflict at Venezuela's Oil Company." Journalist Jean van de Walle says, "Mr. Calderón Berti further jolted Petróleos de Venezuela's management by naming members of the firm's board of directors who were personally loyal to him. Company insiders say the new members couldn't have met the usual criteria of merit and experience that usually governs who goes on the board."

5. Getting hooked on an oil paycheck is not an exclusive trait of developing countries, as the Norwegian experience has clearly shown.

6. General Alfonzo Ravard, in the speech he gave in April 1981 to the students of the Venezuelan Institute of Higher Administrative Studies, defined participatory planning as "that which profits from the information of those who will have to execute the plan so that there is a guarantee that such plans are feasible." In R. Alfonzo Ravard, *Cinco Años de Normalidad Operativa,* p. 366.

7. Even today prominent leaders of moderate political parties mention the necessity "to nationalize the minds" of oil-industry executives. See the comments of Haydeé Lopez Acosta, the COPEI congressional representative quoted in chapter 14.

8. G.W. Grayson, *The Politics of Mexican Oil* (Pittsburgh, Pa.: Pittsburgh University Press, 1980).

9. K. Deutsch, *Politics and Government* (Boston: Houghton Mifflin, 1970). p. 42.

Bibliography

Books

Aberbach, J., R.D. Putnam and B.A. Rockman. *Bureaucrats and Politicians in Western Democracies.* Cambridge, Mass.: Harvard University Press, 1981.

Acosta Hermoso, E. *Petroquímica, Desastre o Realidad?* Caracas, 1977.

Akins, J.E. "The Oil Crisis: This Time the Wolf Is Here." *Foreign Affairs* (April 1973).

Alfonzo Ravard, R. *Cinco Años de Normalidad Operativa, 1975–1980.* Petróleos de Venezuela, Caracas, 1981.

Anastassopoulos, J.P. "The Strategic Autonomy of Government-Controlled Enterprises Operating in a Competitive Economy." Unpublished Ph.D. diss., Columbia University, New York, 1973.

Arnold, R. *The First Big Oil Hunt,* New York: Vantage Press, 1960.

Betancourt, R. *Venezuela, Política y Petróleo* Seix Barral. Caracas, 1978.

Blank, D.E. *Politics in Venezuela,* Boston, 1973.

Bohi, D., and Russel, M. *Limiting Oil Imports: An Economic History and Analysis,* Baltimore: Johns Hopkins University Press, 1978.

Bursk, E., and T. Blodgett, eds., *Developing Executive Leaders.* Cambridge: Harvard University Press, 1971.

Carrera, G.L. *La Novela del Petróleo en Venezuela.* Caracas, 1972.

Coombes, D. *State Enterprise, Business or Politics.* London: Alden and Mowbray, 1971.

Coronel, G. "Recursos de Hidrocarburos en las Areas Inexploradas de Venezuela." *SIVP Bulletin.* Caracas, 1972.

Coronel, G. "Oil towards the Age of Balance.." Annual Report 1981–1982, University of Tulsa, 1982.

Dahl, R.A. *Dilemmas of Pluralist Democracy.* New Haven: Yale University Press, 1982.

Danielsen, A.L. *The Evolution of OPEC* New York: Harcourt, Brace Jovanovich, 1982.

Deutsch, K. *Politics and Government* Boston: Houghton Mifflin, 1970.

Downs, A. *An Economic Theory of Democracy.* New York: Harper & Row, 1975.

Fabrikant, R. "Pertamina, a National Oil Company in a Developing Country," *Texas International Law Journal* 10 (1975).

Fleishman, J.L., L. Liebman, and M. Moore, eds. *Public Duties: The Moral Obligation of Government Officials.* Cambridge: Harvard University Press, 1982.

285

Forbes, R.J., and D.R. O'Beirne. *The Technical Development of the Royal Dutch Shell, 1890–1940.* Leiyden, 1957.

Fundación John Boulton. *Política y Economía en Venezuela, 1810–1976.* Caracas, 1976.

Gil, J.A. *"The Challenges of Venezuelan Democracy,"* New Brunswick, N.J.: 1981.

Gallad, I. Rodriguez, and Yanez, F. *Cronología Ideológica de la Nacionalización Petrolera en Venezuela.* Caracas, 1977.

Grayson, G.W. *The Politics of Mexican Oil.* Pittsburgh, Pa.: Pittsburgh University Press, 1980.

Grayson, L.E. *National Oil Companies.* New York: John Wiley and Sons, 1981.

Hamilton, C.W. "Petroleum Developments in Venezuela during 1930," *American Institute of Mining and Metallurgical Engineers Transactions* 42.

INVEPET. "Diagnóstico sobre Transferencia Tecnológica de la Industria Petrolera," Caracas, 1975.

Kissinger, H. *Years of Upheaval.* Boston: Little, Brown, 1982.

Lagoven. Informe Anual, 1981.

Lieuwen, E. *Petroleum in Venezuela, a History.* London: Russell and Russell, 1967.

Madelin, H. *Oil and Politics.* Lexington, Mass.: D.C. Heath, Lexington Books, 1975.

Maraven. Informe Anual, 1981.

Martínez, A. *Chronology of Venezuelan Oil.* London: Allen and Unwin, 1969.

Martínez, A. *Cronología del Petróleo Venezolano.* Caracas: Ed. Foninves, 1976.

Mazzolini, R. *Government Controlled Enterprises: International Strategic and Policy Decisions.* New York: John Wiley and Sons, 1979.

Mendez, J., and J.R. Domínguez. "La Exploración para Hidrocarburos en Venezuela," Exploration Seminar Maracaibo, 1975.

Ministry of Mines and Hydrocarbons, Memoirs, 1966.

Ministry of Energy and Mines. *"Petróleo y Otros Datos Estadísticos,"* Caracas, 1980, 1981.

Narain, L. *Principles and Practices of Public Enterprise Management.* New Delhi: S. Chand, 1980.

Ninth National Engineering Congress, Transactions, Maracaibo, 1974.

Owen, E. *"Trek of the Oil Finders,"* American Association of Petroleum Geologists Special Publication, Tulsa, Okla., 1975.

Pérez Alfonzo, J.P. *"Oil and Dependency,"* Caracas, 1971.

Petróleos de Venezuela. Informe Anual, 1981.

Philip, G. *Oil and Politics in Latin America.* Cambridge: Cambridge University Press, 1982.

Pick, P. "Letter from Cambridge: Proposal for Reorganization of Venezuelan State-Owned Industry," Harvard University, Center for International Affairs, 1982.

Rabe, S. *The Road to OPEC: United States' Relations with Venezuela, 1919–1976.* Austin: University of Texas Press, 1982.

Sader Pérez, R. *The Venezuelan State Oil Reports to the People.* Caracas, 1969.

Sigmund, P. *Multinationals in Latin America.* Madison: University of Wisconsin Press, 1980.

Smith, P.S. *Oil and Politics in Modern Brazil.* Toronto: McMillan, 1976.

Sullivan, W.M. "The Rise of Despotism in Venezuela: Cipriano Castro, 1899–1908," Ph.D. diss., University of New Mexico, Albuquerque, 1974.

Third Venezuelan Petroleum Congress, Transactions, Caracas, 1974.

Vallenilla, L. *Auge, Declinación y Porvenir del Petróleo en Venezolano.* Caracas, 1973.

Vallenilla, L. *Oil, The Making of a New Economic Order: Venezuelan Oil and OPEC.* New York: McGraw-Hill, 1975.

Veloz, R. *"Economía y Finanzas de Venezuela, 1830–1944,"* Caracas, 1975.

Vernon, R. Harvard Institute for International Development Development Paper 3, Harvard University.

Vernon, R., and Y. Aharoni, eds., *State-Owned Enterprise in the Westtern Economics.* New York: St. Martin's Press, 1981.

Waddams, F.C. *The Libyan Oil Industry.* Baltimore: John Hopkins University Press, 1980.

Williams, E.J. *The Rebirth of the Mexican Petroleum Industry.* Lexington, Mass.: D.C. Heath, Lexington Books, 1979.

Magazines

The most complete source of information and analysis of Venezuelan oil industry events since 1974 has been *Resumen,* published weekly in Caracas. Other magazines used for this book include *Número, Auténtico,* and *Zeta,* all published in Caracas.

Newspapers

The main sources for this book were *El Nacional, El Universal,* and *El Diario de Caracas,* published in Caracas, and *The Wall Street Journal,* and *The New York Times.*

Less frequently used were *Ultimas Noticias,* Caracas paper, and *Panorama,* from Maracaibo.

Other Sources

A Venezuelan monthly report, "The Monthly Report," produced by T. Bottome in Caracas for private circulation, is probably the best source of information on current economic events in Venezuela, including those related to oil-industry activities.

Index

289

About the Author

Gustavo Coronel, a native of Venezuela, was trained as a geologist at the University of Tulsa and worked for 27 years in the oil industry in Venezuela, the Netherlands, the United States and Indonesia. He worked as an exploration geologist, production engineer, refining manager, and trading manager before becoming a member of the board of Petróleos de Venezuela, the holding company of Venezuela's nationalized oil industry and the largest corporation in Latin America. He held this position from 1975 to 1979 and in 1980 was appointed executive vice-president of Meneven, one of the four national oil companies operating in Venezuela. In 1982 he joined the Center for International Affairs at Harvard University.